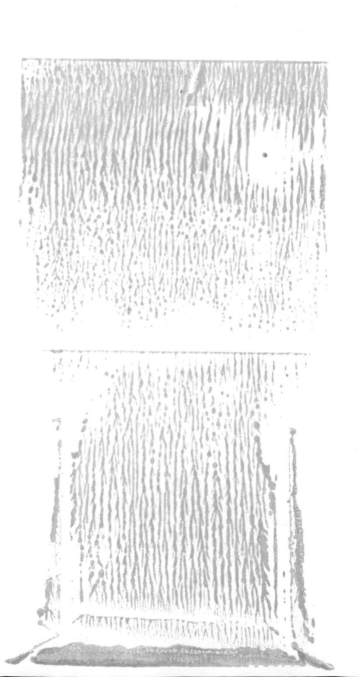

Power Control Electronics

BOYD LARSON

Waukesha County Technical Institute

PRENTICE-HALL, INC., Englewood Cliffs, NJ 07632

Library of Congress Cataloging in Publication Data

Larson, Boyd.
   Power control electronics.

   Includes index.
   1. Power electronics.   I. Title.
TK7881.15.L37   1983   621.31'7   82-15028
ISBN 0-13-687186-0

Editorial/production supervision by Lori Opre
Manufacturing buyer: Gordon Osbourne
Cover design by 20/20 Services, Inc.

Printed in the United States of America

10 9 8 7 6 5 4 3

ISBN-0-13-687186-0

PRENTICE-HALL INTERNATIONAL, INC., *London*
PRENTICE-HALL OF AUSTRALIA PTY. LIMITED, *Sydney*
EDITORA PRENTICE-HALL DO BRASIL, LTDA., *Rio de Janeiro*
PRENTICE-HALL CANADA INC., *Toronto*
PRENTICE-HALL OF INDIA PRIVATE LIMITED, *New Delhi*
PRENTICE-HALL OF JAPAN, INC., *Tokyo*
PRENTICE-HALL OF SOUTHEAST ASIA PTE. LTD., *Singapore*
WHITEHALL BOOKS LIMITED, *Wellington, New Zealand*

Contents

CHAPTER 9   SCR PHASE-CONTROL CIRCUITS   83

CHAPTER 10   THREE-PHASE SCR PHASE CONTROL   101

CHAPTER 11   ELECTRONIC INVERTERS   115

Preface

This text is to be used in a two-year, post-high school, electronics servicing program for courses traditionally called *industrial electronics*. The content of these courses has been inconsistent from school to school: Some schools include electrical machinery, some include digital electronics, some include industrial instrumentation, some include medical electronics, and still others include numerical control of machine tools. Yet nearly all of these courses cover a fairly large amount of material on SCRs and the control of large amounts of electrical power. This text covers only this core material.

The prerequisites are a good knowledge of ac and dc electricity. Some background in electronic devices, electrical machinery, and especially three-phase electrical power is helpful.

A series of 16 experiments written by the author for use with this text are currently available for a nominal fee at the Waukesha County Technical Institute Bookstore, 800 Main Street, Pewaukee, Wisconsin 53072, and should be requested as the laboratory manual for Electronic Motor Control, course number 414 387.

BOYD LARSON

Pewaukee, WI

Introduction to Electronic Power Control

1

This chapter compares the methods of controlling electrical power. The student is presented with the advantages and disadvantages of each. This brief study shows why *transistors* and *silicon controlled rectifiers* (SCRs) are replacing most other methods.

1.1 ELECTRICAL POWER CONTROL METHODS

1.1.1 The Seven Methods of Controlling Electrical Power

The seven general methods of controlling electrical power are:

1. manual switch
2. gas-filled tubes
3. semiconductors
4. rheostat
5. variable reactance
6. variable transformers
7. rotating machinery

The first three methods may be classed as switching methods. Comparing the advantages and disadvantages of the manual switch with the rheostat demonstrates the advantages and disadvantages of all seven methods.

1.1.2 Rheostat Versus Switch

Figure 1.1 shows a rheostat controlling an electric heater. The rheostat has the advantage of controlling the power delivered from nearly zero to full power. At half power, the very inefficient rheostat consumes as much power as the heater. It has the same current through it and an equal voltage across it. Thus, it has equal power.

Besides being inefficient, the rheostat must be physically larger than the heater in order to dissipate additional heat so that it operates at lower temperatures. Using a rheostat to control large amounts of power is very expensive because of energy loss and its construction. The rheostat is heavy, bulky, and difficult to install.

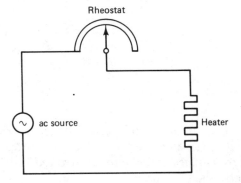

Figure 1.1 A rheostat controlling a heater.

A switch controlling a heater is shown in Fig. 1.2. When the switch is on, it is a short and has no voltage across it. Thus, there is no power dissipated by it. When the switch is off there is no current through it. Again, it has no power loss. The switch does not waste power in either of its two possible states (off or on).

The switch cannot be set at intermediate points as a rheostat can. But, it can produce the same effect by being alternately turned off and on. Transistors and SCRs may be used as a switch. These devices may be automatically turned off and on hundreds of times a second. When more heat is desired, the electronic switch is set on for longer periods than it is set off. For less heat, it is set off longer.

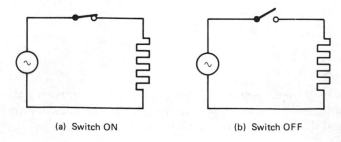

(a) Switch ON   (b) Switch OFF

Figure 1.2 A switch controlling a heater.

1.1.3 Variable Reactance Controllers

Variable reactors yield proportional control with low power losses. The electrical property of inductive reactance does not consume power, but it is an opposition to current. The primary methods of controlling power with reactance are:

1. switching a *tapped inductor*
2. using a *saturable reactor*

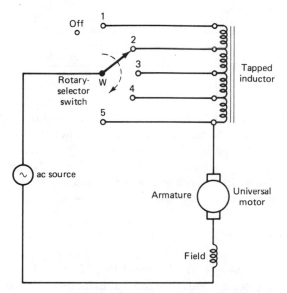

Figure 1.3 A tapped inductor controlling the speed of a universal motor.

Figure 1.3 shows a tapped inductor being used to control the speed of a universal motor. With the wiper (*W*) of the rotary-selector switch in position 1, current from the source must flow through the full inductance. This is the slowest speed, excluding off. More inductance is switched out of the path at higher speed positions. In position 5, all of the inductance is out of the path and the motor runs at full speed. Tapped inductors are used in many two-speed small appliances.

Figure 1.4 shows a saturable reactor used as a lamp dimmer. The saturable reactor has two windings on a common iron core. The dc winding is used as the reactance controller. This winding has many turns of fine wire. A small current through it yields a *flux* density large enough to saturate the iron core. When the core is saturated its reactance is small and the ac winding nearly becomes a short. This allows a high current through the lamp. Making the core less saturated causes the ac winding to have more reactance, limiting current through the lamp. The ac winding has only a few turns of heavy wire. The wire must be heavy enough to support load current.

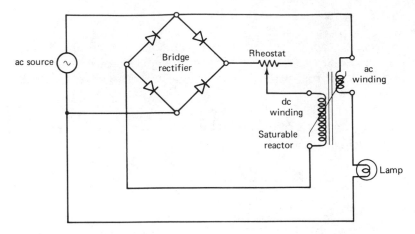

Figure 1.4 Saturable reactor as a lamp dimmer.

The principle of the saturable reactor is a small dc current controlling the reactance of a large ac winding. Saturable reactors are actually amplifiers, as a low dc power controls a large ac power. Some rectify a portion of their output current and feed it into a second dc winding, aiding the primary dc winding. This is called *regenerative feedback*. A saturable reactor with regenerative feedback is called a *magnetic amplifier*.

1.1.4 Variable Autotransformers

Variable autotransformers are very efficient machines. Figure 1.5 shows the schematic and pictorial sketch of a variable autotransformer. The single coil is wound on a donut-shaped iron core. The variable tap operates like a wire-wound potentiometer wiper. These transformers have a full range of control from zero to full power. At zero power a small primary current flows but a switch may be used to turn it off.

The large and bulky variable autotransformer is expensive to construct. It is fairly popular for small loads.

1.1.5 Gas-Filled Electron Tubes

Gas-filled tubes are the forerunners of semiconductor-switching devices. The smaller capacity gas-filled tubes are called *thyratrons*. These are still being used in some older equipment. *Ignitrons* are being designed for new equipment. They have a much larger single-device current capacity than the largest SCRs and can withstand overloads better. (Chapter 19 of *Transistor Fundamentals and Servicing*, Larson, Prentice-Hall, Inc., 1974 covers these devices more thoroughly.)

4

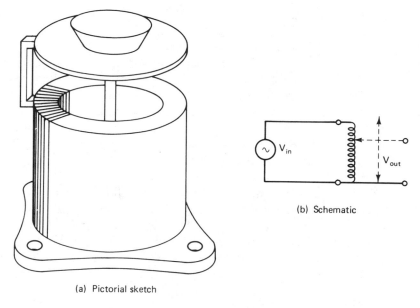

(b) Schematic

(a) Pictorial sketch

Figure 1.5 The variable autotransformer.

Gas-filled tubes, transistors, and SCRs are all electronic-switching devices. They operate on the principle of the switch. The switch controls power by varying the time the load is turned on. Electronic switches operate many times a second. Their fast action delivers variable power as smoothly as a rheostat and as efficiently as a switch.

1.2 ELECTRONIC SWITCH-CONTROL METHODS

1.2.1 The General Methods of Electronic-Switching Control

The four general methods of varying electric power with electronic switches are:

1. phase control
2. zero-voltage switching
3. inverter control
4. static switching

The first three methods are demonstrated by their waveforms and path of principal current. The significant point in this chapter is to differentiate between the four methods. Each will be covered in separate chapters later in the text.

1.2.2 Phase Control

Figure 1.6(a) is a phase-control circuit using a full-wave bridge and an SCR to control an ac load. Figure 1.6(b) is identical except for the load connection. The load is connected into the dc side of the bridge instead of the ac. This shows that the circuit may be used to control either ac or dc loads.

The SCR is turned on when its anode and gate are both positive. It is turned off only when its anode voltage goes to zero. The full-wave bridge supplies the SCR with an unfiltered dc voltage V_{dc} [see Fig. 1.6(d)]. V_{dc} goes to zero twice on every incoming cycle. This automatically turns the SCR off 120 times per second on a 60-Hz line.

The SCR is off with no gate-trigger voltage. Then no current can flow through it or the load. Hence, the load is turned off. In order to turn the load fully on, the gate trigger must turn on (fire) the SCR at the start of every cycle of V_{dc}. The *gate-trigger voltage* (V_t) must be synchronized with V_{dc}. In Fig. 1.6(e), V_t is shown at 60° every time V_{dc} goes through zero. This fires the SCR at 60° and 240°, as shown by the SCR current waveform (I_{SCR}) in Fig. 1.6(f).

The voltage across the SCR is V_{dc} until the SCR is fired [see Fig. 1.6(g)]. The ON voltage across the SCR varies from 0.7 to 1.5 V, but is shown as zero in Fig. 1.6(g). When the SCR is on, the main voltage drop is across the load, but zero *IR* drop across the load when the SCR is off [see Fig. 1.6(h) and (i)].

Lower power is delivered by delaying the gate trigger. If the gate trigger were on at 90°, only half power would be delivered to the load. If it were on at 170°, almost no power would be delivered.

The trademark of phase control is the "chopped"-half sine wave [see Fig. 1.6(f), (g), (h), or (i)]. Its advantage is simplicity and high efficiency, although the examples shown in Fig. 1.6(a) and (b) are the most inefficient phase-control circuits because of the drop across two diodes and the SCR. This circuit is chosen for its versatility. That is, it is capable of controlling either an ac or dc load.

Phase control is suited for 60 Hz, but not higher frequencies. The main disadvantage of phase control is *electromagnetic interference* (EMI), sometimes known as *radio-frequency interference* (RFI). The chopped-half sine wave emits strong *harmonics*. They interfere with radio, television, and other equipment. One method of arresting RFI is with shielding and filters. Another method is by using *zero-voltage switching*.

1.2.3 Zero-Voltage Switching

In order to avoid the chopped-half sine wave associated with phase control, zero-voltage switching may be used. The output of zero-voltage switching is shown in Fig. 1.7.

6

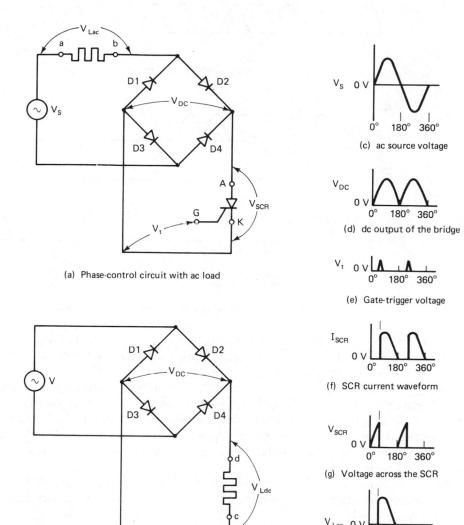

(a) Phase-control circuit with ac load

(b) Phase-control circuit with dc load

(c) ac source voltage

(d) dc output of the bridge

(e) Gate-trigger voltage

(f) SCR current waveform

(g) Voltage across the SCR

(h) Voltage across the ac load

(i) Voltage across the dc load

Figure 1.6 An SCR phase-control circuit with waveforms.

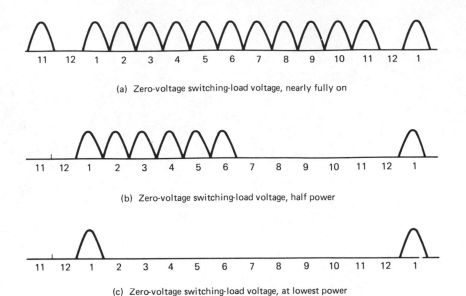

(a) Zero-voltage switching-load voltage, nearly fully on

(b) Zero-voltage switching-load voltage, half power

(c) Zero-voltage switching-load voltage, at lowest power greater than fully off

Figure 1.7 Zero-voltage switching output voltages.

Figure 1.7(a) shows nearly full-power load voltage. Figure 1.7(b) shows a load voltage set at half power. Figure 1.7(c) shows a zero-voltage switching-load voltage set at a very low power.

The disadvantage of zero-voltage switching is that the switching may be observable on low-frequency sources. A lamp dimmer on 60 Hz (120-Hz rectified dc) would flicker. The human eye can detect flicker lower than 16 times per second. The setting shown in Fig. 1.7(c) would flicker once every twelve cycles. At 120 Hz this is 10 times per second. This is highly visible and annoying.

Zero-voltage switching lends itself to frequencies higher than 60 Hz. In order to attain high power at higher frequencies, a *cycloconverter* is generally used.

The advantages of zero voltage switching are twofold. First, it suppresses annoying and sometimes dangerous RFI. Second, it suppresses high-voltage transients encountered with inductive loads. Transients are dangerous to electronic and electrical devices and circuits. Zero-voltage switching will be covered further in Chapter 12.

1.2.4 Electronic-Inverter Switching

Electronic-switching circuits may be used to replace *motor-generator* (MG) sets, such as inverters, converters, and cycloconverters. Electronic switches used in converters and cycloconverters are said to be used in *inverter service*. Inverter type SCRs have faster switching and higher frequency capabilities.

An inverter changes dc to ac. A *dc converter* changes one dc voltage to a different dc voltage. An *ac converter* changes ac from one voltage to another. A cycloconverter changes ac from one frequency to another.

Inverters are used in applications requiring portable ac power and in cycloconverters. An electronic cycloconverter consists of a rectifier followed by an inverter.

One of the fastest growing applications for cycloconverters is *induction motor-speed control*. Induction motor speed is controlled by the frequency of the source. Motor-speed control was formerly dominated by dc motor systems, where speed control required only a variable dc source. Ac induction motor control has grown more popular since the advent of high-power, high-speed electronic switches.

Figure 1.8 is a schematic of a single-phase inverter. A square-wave generator and transformer T_1 supplies Q_1 and Q_2 with out-of-phase inputs. These switch Q_1 and Q_2 on and off sequentially. When Q_1 is on, current flows from the +12 V_{dc} source through the top half of the primary of T_2. This induces the top half of the 190 V square wave across the secondary of T_2. Then, Q_2 conducts allowing current through the bottom of T_2. This induces the negative portion of the 190 V square-wave output. The *Ott filter* converts the square wave to a sine wave.

A sine-wave generator could be used in place of the square wave. The transistors could be biased to amplify it. Then the Ott filter could be eliminated. However, at 30°, 120°, 210°, and 300° the transistors would be half on. They would drop half of the supply voltage and be as inefficient as the rheostat shown in Fig. 1.1. The transistors would be operating like a rheo-

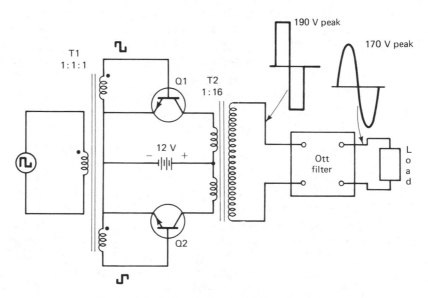

Figure 1.8 A single-phase transistor inverter.

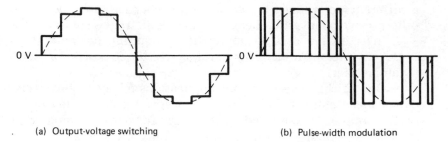

(a) Output-voltage switching   (b) Pulse-width modulation

Figure 1.9 Sine-wave synthesis with rectangular pulses.

stat, not a switch. The transistors operate with the efficiency of a switch when the square wave is used.

Figure 1.9 shows two commonly used inverter outputs. The solid lines outline rectangular waveforms which are typical inverter outputs. The dotted lines show how the rectangular pulses approximate a sine wave. Inverter circuits are covered in more detail in Chapter 11.

1.2.5 Static Switching

SCRs and a few other types of electronic power devices are used as *static switches*. These applications replace manual on/off switches, relays, circuit breakers, fuses, and flashers. Some common uses are switching on emergency lighting during blackouts, and in flashers on portable roadside warning lights.

These switches can easily be actuated automatically. They are inexpensive and reliable. They may be used to actuate ac or dc circuits.

The term *static* means not changing. In this respect the static switch turns a power off or on, but does not vary it in other ways. The term static excludes proportional controller where the percentage of fully on is varied. The on/off changes are constant, even in the case of flashers. Static-switching circuits are covered in more detail in Chapter 13.

1.2.6 Summary of Power Electronics

There are several methods of varying power: the manual switch, rheostat, reactors, transformers, and electronic switches. The manual switch is the cheapest and most efficient. Electronic switches have the advantage of being efficient and may be used to vary power automatically.

There are three electronic-switching methods of delivering variable electric power: phase control, zero-voltage switching, and inverters. These three methods may be differentiated by their output waveforms (see Fig. 1.10).

A fourth type of electronic-switching control is called static switching. This type uses the electronic switch to replace manual switches and circuit breakers.

(a) Phase-control output waveform

(b) Zero-voltage switching-output waveform

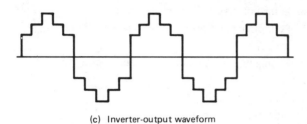

(c) Inverter-output waveform

Figure 1.10 Output waveforms of the three electronic-switching methods of controlling electric power.

The schematics in this chapter are single-phase circuits with the gate trigger omitted. Later chapters cover the gate-trigger circuits and three-phase circuitry.

PROBLEMS

1.1. What two devices are replacing other methods of controlling electric power?

1.2. List seven methods of controlling electric power.

1.3. Find the power consumed by a rheostat set at 2 Ω and controlling a 1.5-Ω heater across a 480-V line.

1.4. Find the power consumer by an ON switch controlling a 100-A heater on a 240-V_{ac} line.

1.5. Find the power consumed by an OFF switch controlling a 6-kW heater on a 240-V_{ac} line.

1.6. What is the advantage of the rheostat over the manual switch as a power controller?

1.7. What is the advantage of the manual switch over the rheostat as a power controller?

1.8. List two types of reactors which are used in power control.

1.9. What is the advantage of the variable-reactor controller over the rheostat?

1.10. What type of reactive controller distorts the ac sine wave?

1.11. Sketch the schematic of a variable autotransformer.

1.12. Which type of gas-filled tube is not being used in new designs?

1.13. Which electronic-switching device has the largest single-device capability?

1.14. List three methods of varying electric power with electronic switches.

1.15. Which type of load is the circuit shown in Fig. 1.6 capable of controlling (ac, dc, or both)?

1.16. What voltage polarity must be on the anode with respect to the cathode for the SCR to be fired?

1.17. What voltage polarity must be on the gate with respect to the cathode for the SCR to be fired?

1.18. How many times per second is an SCR turned off when it is operated with phase control on a 50-Hz line in the circuit shown in Fig. 1.6?

1.19. What is the trademark of the phase-control method of controlling power?

1.20. What is the main disadvantage of the phase-control method?

1.21. What is the main disadvantage of zero-voltage switching?

1.22. What method of control eliminates the problem of excessive RFI?

1.23. What circuit changes dc to ac?

1.24. What is the advantage of using high frequencies with zero-voltage switching?

1.25. List three uses of inverters.

1.26. How is the speed of a dc motor controlled?

1.27. How is the speed of an induction motor controlled?

1.28. What is an Ott filter used for?

1.29. Why do inverter circuits use rectangular waveforms instead of sine waves?

1.30. Sketch two methods of synthesizing a sine wave using rectangular waves.

1.31. List four applications of static switching.

1.32. Sketch the output of a zero-voltage switching circuit. Show the period for six rectified dc pulses with the SCR delivering half power to the load.

Single-Phase Power Rectifiers

2

This chapter is a review of single-phase rectifier circuits. Later in the text the same circuits are presented with diode/SCR combinations, with SCRs only, and again only reversed with SCRs or transistors as inverters.

A rectifier converts readily available ac power to dc. The rectifier does this electronically without using any moving parts. The most popular rectifying device is the *semiconductor diode.*

2.1 THE SEMICONDUCTOR DIODE

2.1.1 The Basic Diode

The schematic of the semiconductor diode is shown in Fig. 2.1. The diode has two elements—an *anode* and a *cathode*. When the anode is more positive than the cathode, the diode acts like an ON switch. That is, the diode conducts current readily. When the cathode is more positive than the anode, the diode acts like an OFF switch.

Diodes are made of two materials: silicon and germanium. Power-rectifier diodes are usually made of silicon. Silicon has the ability to withstand higher temperatures and has a higher reverse resistance.

Figure 2.2 shows the diode characteristic curve. This is a plot of the set of volt-ampere points. Voltage is shown on the horizontal axis, and current on the vertical. Both axes show negative and positive values of voltage and current. The diode begins to conduct as the voltage across its anode, with

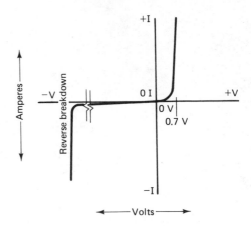

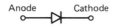

Figure 2.1 The schematic symbol of a semiconductor diode.

Figure 2.2 The characteristic curve of a semiconductor diode.

respect to its cathode, is increased in the positive direction. At about 0.7 V, in silicon diodes any slight increase in voltage greatly increases current. This increase in current is nearly without bound. Hence, diodes used with voltages over 0.7 V (which covers most diode applications) must have a resistance in series with it to limit current through it.

When the voltage from anode to cathode is increased in the negative direction, the diode acts like an open switch. Actually, a diode may have from 1 to 1000 MΩ of resistance in this reverse direction. This large-resistance characteristic continues as voltage is increased (negatively) until reverse-breakdown voltage (see Fig. 2.2). At this value, a diode will turn on in the reverse direction. Again, if current is not limited by a series resistance the current will increase without bound and destroy the diode.

The *peak inverse voltage* (PIV) of a diode is the manufacturer's rating of a particular diode. The PIV is the largest reverse voltage a diode should be subjected to. Some common PIV ratings are 100, 200, 300, 400, and 600 V.

Figure 2.3(a) shows a *forward-biased diode* and Fig. 2.3(b) shows its switch-equivalent circuit. When a silicon-diode anode is more than 0.7 V positive than its cathode, it is considered to be forward-biased. It may be considered to act like an ON switch. Figure 2.3(c) shows a *reverse-biased diode* and Fig. 2.3(d) shows its switch equivalent. A reverse-biased diode has its anode more negative than its cathode and may be considered an OFF switch.

The diode-switch equivalent circuits shown in Fig. 2.3(b) and (d) are useful in analyzing diode circuits and describing how diodes may be tested. Power-rectifier diodes may easily and safely be tested with an ohmmeter. Connect the ohmmeter so that its internal dry cell forward-biases the diode. This should yield a low-resistance reading. Reversing the leads should yield a much higher resistance or even an infinite reading. Precautions should be taken when testing low current or low PIV diodes in this manner, as some ohmmeters may put out voltages and/or currents which exceed a diodes rating (see Sec. 4-17 in *Transistor Fundamentals and Servicing*, Larson, Prentice-Hall, Inc., 1974).

2.1.2 Diode Circuit Analysis

There are two equivalent diode circuits which may be used to mentally simplify diode circuit analysis. One is the switch equivalent shown in Fig. 2.3.

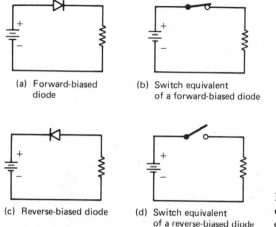

(a) Forward-biased
    diode

(b) Switch equivalent
    of a forward-biased diode

(c) Reverse-biased diode

(d) Switch equivalent
    of a reverse-biased diode

Figure 2.3 The two states of a diode—forward or on, and reverse or off.

This is probably the most valuable one. However, in some circuits it may be difficult to determine which switch equivalent to apply (OFF or ON)—for example, in circuits with more than one source or with more than one diode in series. In these circuits it is helpful to mentally replace the diode(s) by resistances. In either *state* (OFF or ON) any good diode has some resistance.

2.2 SINGLE-PHASE RECTIFIER CIRCUITS

2.2.1 The Half-Wave Rectifier

Figure 2.4 shows a silicon diode in an ac circuit. When the ac voltage at terminal A is 0.7 V with respect to ground, the diode turns ON. This allows current through resistor R. Thus, the voltage drop across R (V_o) follows the positive half-sine wave for all voltages greater than 0.7. When the voltage at A falls below +0.7, the diode turns OFF. Then, no current flows through R. The voltage output across R (V_o) is shown in Fig. 2.4(e).

The circuit in Fig. 2.4(a) is called a half-wave rectifier. It changes ac power to dc. Its output is pulsating dc, but it may be used for many high-power loads.

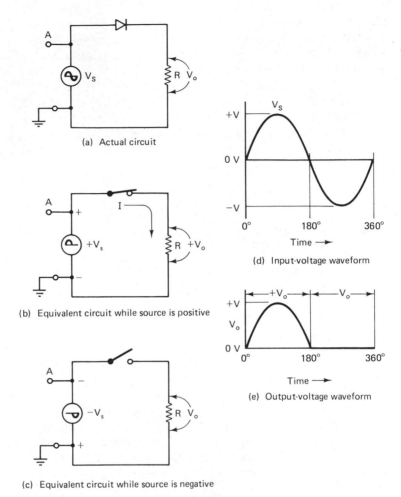

(a) Actual circuit

(b) Equivalent circuit while source is positive

(c) Equivalent circuit while source is negative

(d) Input-voltage waveform

(e) Output-voltage waveform

Figure 2.4 Half-wave rectifier shown with the equivalent circuits—while source is positive and while it is negative.

2.2.2 Diode Circuit Analysis with an ac Source

Ac circuits have a time-varying voltage. Hence, there may be times when the voltage forward-biases a diode and times when it reverse-biases the same diode. The analysis must be divided into two parts. It must be noted at which times the voltage polarity across the diode results in forward-biasing and which times it offers reverse-biasing. The diode may be mentally replaced by an OFF (open) switch when it is reverse-biased, and by an ON (closed) switch when it is forward-biased.

Example 2-1

Find the ac voltage across R (V_o) in Fig. 2.4 during one cycle of the ac source voltage (V_s) as shown in Fig. 2.4(d).

Solution

1. *Determine when the diode is forward-biased.* When the top of the source in Fig. 2.4(a) goes positive, the anode is more positive than its cathode and the diode is forward-biased.

2. *Redraw the circuit replacing the diode with an ON switch.* The voltage across R in Fig. 2.4(b) is identical to the voltage of the source shown in Fig. 2.4(d) for the first $180°$ (assuming the 0.7 V turn-on voltage is trivial).

3. *Determine when the diode is reverse-biased.* When the source is negative the anode of the diode is more negative than its cathode and the diode is reverse-biased.

4. *Redraw the circuit replacing the diode with an OFF switch.* The negative half of the source is dropped across the diode leaving zero volts across the resistor R [see Fig. 2.4(e)] from $180°$ to $360°$.

2.2.3 The Full-Wave Centertapped-Transformer Rectifier

Figure 2.5(a) shows the schematic diagram of the full-wave centertapped-transformer rectifier. When terminal A end of the transformer is positive, diode D_1 is ON [see Fig. 2.5(b)] and current flows through R (the load) from O to B causing a positive drop. When terminal C end of the transformer is positive, diode D_2 is ON [see Fig. 2.5(c)] and current flows through R again from O to B causing another positive drop $180°$ to $360°$ [see Fig. 2.5(d)].

2.2.4 The Full-Wave Bridge Rectifier

A second full-wave rectifier is shown in Fig. 2.6(a). This full-wave bridge rectifier uses four diodes. In order to find the state of these diodes, replace them mentally with resistances [see Fig. 2.6(b)]. During the positive half of the source voltage, the drops across the equivalent resistances can be determined. The polarity of the drops found in Fig. 2.6(b) may then be applied to the diodes in Fig. 2.6(a). Thus, the state of each diode may be determined as shown in Fig. 2.6(c) by the diodes-switch equivalent circuit. Current flow during the positive half of the source is through D_2 and the load R, from right to left, and then through D_3 and back to the bottom end of the source. This causes a positive drop across R as shown in Fig. 2.6(c).

Figure 2.7 shows the full-wave bridge circuits during the negative portion of the source voltage. Again, the diodes are replaced mentally with resistors in order to determine voltage polarity [see Fig. 2.7(a)]. Thus, the diodes state may be determined as shown in Fig. 2.7(b). The current path is now through D_4, from right to left through R, and then through D_1 to the source. The current path through R is the same direction as when the source is positive. This causes a positive drop across R during both half cycles [see the waveform in Fig. 2.7(c)].

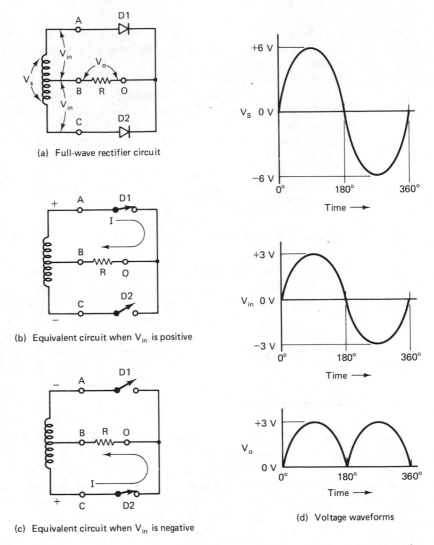

(a) Full-wave rectifier circuit

(b) Equivalent circuit when V_{in} is positive

(c) Equivalent circuit when V_{in} is negative

(d) Voltage waveforms

Figure 2.5 The full-wave rectifier shown with its equivalent circuits and voltage waveforms.

2.2.5 Capacitive Filters

All electronic rectifiers have a pulsating voltage output. A popular method used to decrease these pulsations is the capacitive filter. A capacitor connected across the load tends to greatly reduce these pulsations. Filter capacitors are usually electrolytic in order to get a higher value of capacitance in the smallest package. Larger values of capacitance filter better.

Figure 2.8(a) shows a half-wave rectifier with a capacitive filter. When

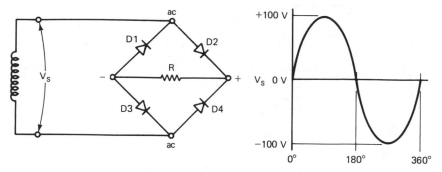

(a) Full-wave bridge rectifier and its input voltage

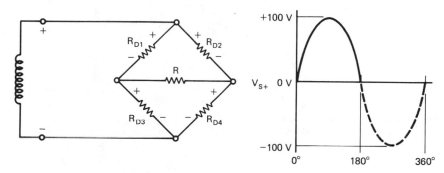

(b) Diodes replaced mentally by resistors to show polarity during the positive half of the input voltage

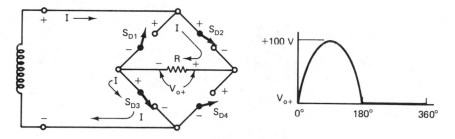

(c) Diodes replaced mentally by switch equivalents to show current flow during the positive half of the input voltage

Figure 2.6 The full-wave bridge-rectifier analysis of its output during the positive half of input voltage.

the source at point A is positive, diode D is forward-biased and turned ON [see Fig. 2.8(b)]. Current flows through the load R and into capacitor C. The diode switch actually connects the source directly across C. Thus, during the first 90° of the sine wave, C charges to the peak voltage of the source [see Fig. 2.8(d) and (e)]. The diode becomes reverse-biased after the source

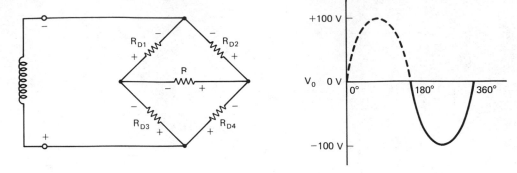

(a) Diodes replaced mentally by resistors to show polarity during the negative half of the input voltage

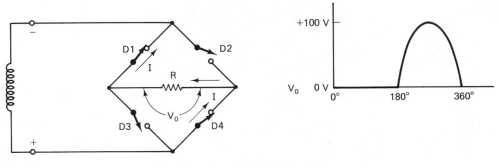

(b) Diodes replaced mentally by switch equivalents to show current flow during the positive half of the input voltage

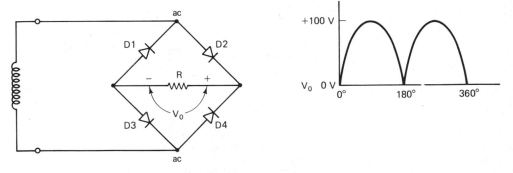

(c) Full-wave bridge rectifier and unfiltered output voltage

Figure 2.7 Analysis of a bridge-rectifier output during the negative half cycle and its total output.

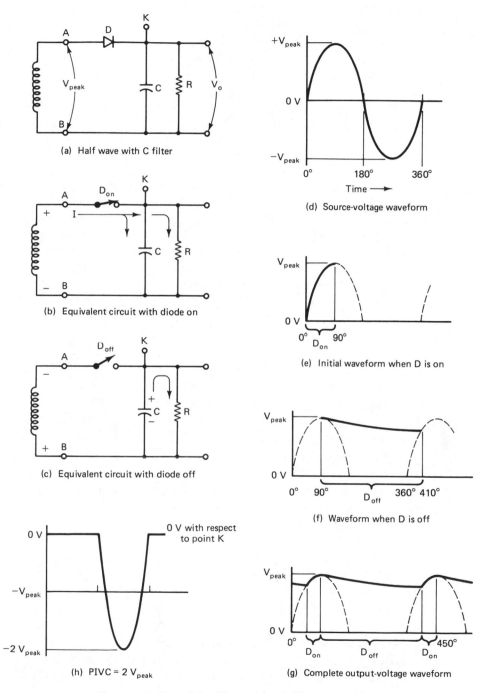

(a) Half wave with C filter

(b) Equivalent circuit with diode on

(c) Equivalent circuit with diode off

(d) Source-voltage waveform

(e) Initial waveform when D is on

(f) Waveform when D is off

(g) Complete output-voltage waveform

(h) PIVC = 2 V_{peak}

Figure 2.8 Capacitive filter with a half-wave rectifier.

peaks and starts going less positive. The positive voltage across the capacitor exceeds the positive voltage across the source. This turns the diode OFF at about $92°$ [see Fig. 2.8(c)]. Then, the capacitor discharges and supplies the load with current during the remaining portion of the cycle [see Fig. 2.8(f)]. The capacitor discharges only a small value of its voltage before the source goes to its positive peak and turns D ON, recharging C to peak voltage.

The RC path for the capacitor charge has a very short *time constant* compared to its discharge path. Both paths contain the same value of C, but the charge path has only the low value of the forward resistance of the diode and ac source internal resistance. Thus, the output waveform is as shown in Fig. 2.8(g).

Figure 2.8(h) shows the peak inverse voltage (PIV) which the circuit presents to the diode (PIVC). The PIVC that the unfiltered half-wave rectifier presents to the diode is only the peak value of the source voltage. This may be seen by looking at the peak voltage across the diode when it is OFF [see Fig. 2.4(c)]. Figure 2.8(c) shows the diode when it is OFF in the filtered half-wave rectifier. The capacitor is charged to nearly the peak value of the source. This voltage is in series with the negative peak of the source. Thus, the PIVC of this circuit is nearly twice the peak value of the source.

PROBLEMS

2.1. Define *rectifier*.

2.2. What is the most popular rectifying device?

2.3. Which diode element is most positive when the diode is ON?

2.4. What type of materials are power rectifiers made of?

2.5. List two advantages of silicon over germanium.

2.6. Define PIV and PIVC.

2.7. What is the minimum voltage across a forward-biased silicon diode?

2.8. Sketch the switch-equivalent circuit of a forward-biased diode.

2.9. Sketch the switch-equivalent circuit of a reverse-biased diode.

2.10. Describe how a diode may be tested using an ohmmeter.

2.11. What diode-equivalent circuit should be used in circuits with two diodes in series in order to analyze the circuit?

2.12. Sketch the schematic of a half-wave rectifier.

2.13. What would be the peak value of voltage of V_o in Fig. 2.5(d) if V_s in Fig. 2.5(a) and (d) is 170-V peak?

2.14. Sketch the output of Fig. 2.6 with all diodes reversed.

2.15. What is the capacitor used for in a rectifier circuit?

2.16. Find the PIVC of the capacitor-filtered half-wave circuit with 120-V_{rms} input.

2.17. Explain why diode D in Fig. 2.8 turns off slightly after the source voltage drops from its peak, instead of turning off at zero volts.

Three-Phase Rectifiers

3

3.1 INTRODUCTION TO THREE-PHASE RECTIFIERS

Three-phase rectifiers are preferred over single-phase in most applications requiring large amounts of power. The same circuits covered in this chapter are used with SCRs and again in inverter circuits (see Chapters 10 and 11).

Dc power is used in industrial battery chargers, electroplating, electric vehicles such as fork lifts, automobiles, golf carts, and overhead cranes, and in X-ray machines.

The advantages of using three-phase rectifiers include all of the advantages of using a three-phase system. Three-phase alternators are more efficient and deliver a constant power output. This gives its prime mover a constant load. Three-phase distribution systems are more efficient and use less copper than single-phase systems. Unfiltered three-phase rectifiers have a smoother dc output. Its pulsations are closer together and less than half of the peak voltage output. Unfiltered single-phase pulsations are always equal to their peak voltage output. Three-phase rectifier circuits use either three, six, or twelve *cells* (diodes or SCRs). More cells distribute the load allowing smaller cells to be used, yielding a lower cost.

3.1.1 Three-Phase Sources

A three-phase source is three separate single-phase alternators mechanically connected so that each voltage is displaced 120° from the other two. The three-phase voltage waveforms are shown as V_{in} in Fig. 3.1(b). The three common methods of connecting three-phase sources are *delta, three-wire wye*, and *four-wire wye*.

The four-wire wye connection is shown in Fig. 3.1(a). This yields three voltages V_A, V_B, and V_C in Fig. 3.1(b). These three voltages are all taken with respect to the neutral connection N. The three-wire wye connection is shown in Fig. 3.3. These voltages are across each pair of the external terminals A, B, or C. The voltages are then V_{AB}, V_{BC}, and V_{CA} and are all 120° with respect to each other. The three-wire wye voltages are all taken across two secondaries. Thus, their magnitude is 1.732 ($\sqrt{3}$) times as large as the four-wire wye or delta voltages, using identical sources.

The delta connection is shown in Fig. 3.2(a). Its voltages are shown in Fig. 3.2(b).

3.2 THE HALF-WAVE THREE-PHASE RECTIFIER

The half-wave three-phase rectifier is shown in Fig. 3.1(a). This circuit is analyzed by first finding the periods each diode is on and then applying its respective source across the load R. This yields the composite output voltage (V_o) shown in Fig. 3.1(b).

The most positive voltage turns its respective diode on. The ON diode switches its respective-most positive source to the other two diode cathodes. This holds the other two diodes off. Thus, only one diode can be on at one time (disregarding the instant of switching).

The graph in Fig. 3.1(a) of V_{in} is used to determine the periods each diode is on. From 0° to 30°, V_C is the most positive voltage. This biases

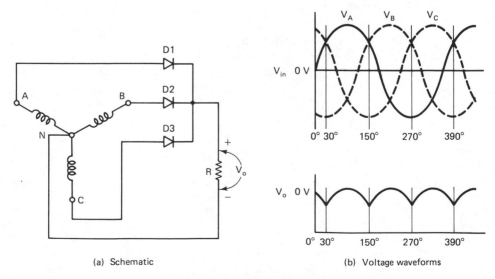

(a) Schematic   (b) Voltage waveforms

Figure 3.1 The half-wave three-phase dc power supply.

diode D_3 on and applies V_C across R as V_o. From 30° to 150°, the most positive voltage V_A forward-biases D_1 and is applied across R as V_o. At 150°, V_B becomes the most positive and switches D_2 on. This applies V_B across R as V_o [see Fig. 3.1(b)] from 150° to 270°. At 270°, V_C becomes most positive and D_3 turns on. Diode D_3 connects V_C across R and it is V_o from 270° to 390°.

The output (V_o) repeats itself three times every 360° of one source cycle. The frequency of the output is three times one of the input frequencies.

3.3 THREE-PHASE FULL-WAVE RECTIFIERS

Figures 3.2 and 3.3 show schematics of three-phase full-wave rectifiers. The analysis of both circuits is identical. The only difference in the two circuits is the three-phase source. One is wye and one is delta. Using the same primary source, the wye voltages would be $\sqrt{3}$ larger and shifted 30° from the delta inputs. However, if V_{in} in Fig. 3.1(b) is considered to be the source voltage of either, the waveform analysis of both circuits are identical.

The three-phase full-wave rectifiers utilize both the positive and negative half of the input as in the single-phase full-wave. The negative peaks are

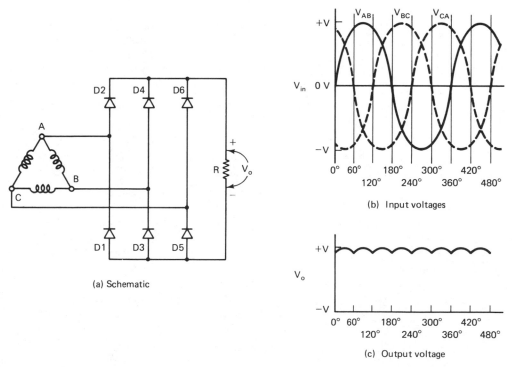

(b) Input voltages

(a) Schematic

(c) Output voltage

Figure 3.2 The full-wave three-phase power supply with a delta source.

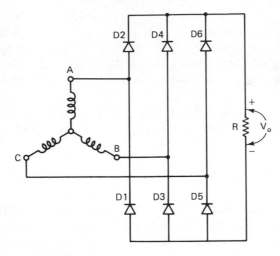

Figure 3.3 The full-wave three-phase dc power supply using a wye source.

flipped over with respect to the load (R). Thus, the frequency of the output is six times the frequency of one of the input voltages [see Fig. 3.2(c)].

3.3.1 Analysis of the Full-Wave Three-Phase Rectifier

Three premises are used in analyzing the operation of the three-phase full-wave rectifier:

1. Two diodes are always on, while the remaining four are off (excluding the instant switching occurs).
2. One of the ON diodes is an even-numbered diode (D_2, D_4, or D_6) and one is always odd (D_1, D_3, or D_5).
3. Current always flows out of the source terminal from the highest positive voltage, through an even-numbered diode, through the load, through an odd-numbered diode, and then back into the source terminal with the highest negative voltage.

Thus, the current path at any time may be found by finding the most positive source terminal and the most negative. The most positive source terminal forward-biases its respective even-numbered diode, turning it on. The most negative source terminal forward-biases its respective odd-numbered diode, turning it on [see Fig. 3.4(a) and (b)].

All that remains is finding which source terminal is most positive and which is most negative. This may be done by graphing two voltages as shown in Fig. 3.5(b). The voltages are plotted with respect to a single common reference terminal. Terminal B is arbitrarily selected as the common reference terminal. Thus, the two voltages are V_{AB} and V_{CB}. V_{AB} is shown in

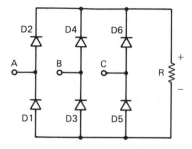

(a) Simplified schematic of either three-phase
full-wave rectifier

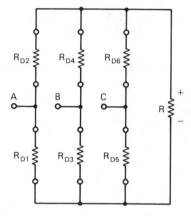

(b) Resistance-equivalent circuit of the full-wave
three-phase rectifier

Figure 3.4 Analysis of the three-phase full-wave rectifiers shown in Figs. 3.2(a) and 3.3.

Fig. 3.5(a) and V_{CB} is graphed in Fig. 3.5(b) as the inverse of V_{BC} as shown in Fig. 3.5(a).

The graph shown in Fig. 3.5(b) is actually the graphs of the three terminal voltages. The voltage at terminal A is actually V_{AB}. The voltage at terminal C is V_{CB}. The voltage at terminal B is represented by the zero volt reference line.

Thus, the state of the diodes may easily be determined from Fig. 3.5(b). Terminal C voltage is highest from $0°$ to $60°$. Therefore, from Fig. 3.4(a), the anode of D_6 is at the most positive voltage in the circuit. This forward-biases D_6 and turns it on. Figure 3.5(b) shows that terminal A is most positive from $60°$ to $180°$. Hence, D_2 is on during this period. At $180°$ terminal A voltage goes below the zero reference line or terminal B voltage. Then, the most positive voltage is at B turning on D_4 [see Fig. 3.4(a)]. At $300°$ the voltage at terminal C goes more positive than B turning D_6 on again.

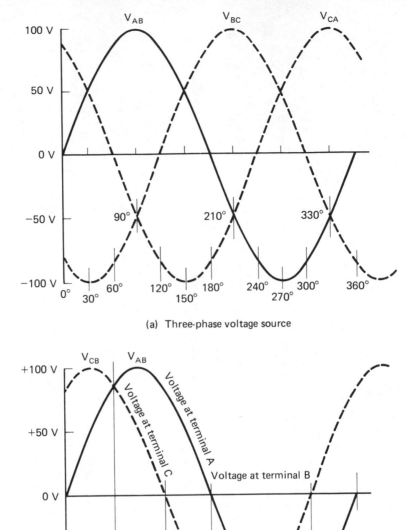

(a) Three-phase voltage source

(b) Voltages at A and C with respect to point B at zero volts

Figure 3.5 Waveform analysis of the three-phase full-wave rectifier.

The ON odd-numbered diode is found by finding the most negative terminal voltage from Fig. 3.5(b). For example, B is most negative from $0°$ to $120°$, C from $120°$ to $240°$, and terminal A is most negative from $240°$ to $360°$. These results are summarized in Fig. 3.6(a). The switch-equivalent

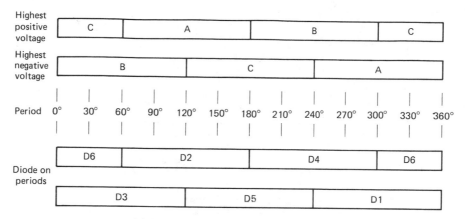

| Highest positive voltage | C | A | B | C |
| Highest negative voltage | B | C | A |

| Period | 0° | 30° | 60° | 90° | 120° | 150° | 180° | 210° | 240° | 270° | 300° | 330° | 360° |

| Diode on periods | D6 | D2 | D4 | D6 |
| | D3 | D5 | D1 |

(a) Chart showing periods of highest positive and
negative voltages and their respective ON diodes

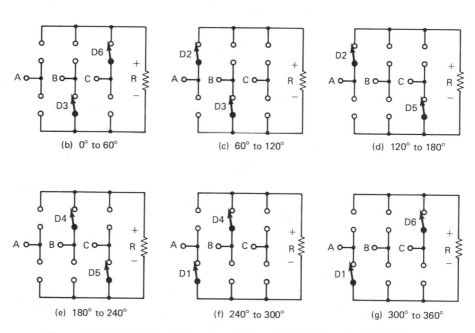

(b) 0° to 60°   (c) 60° to 120°   (d) 120° to 180°

(e) 180° to 240°   (f) 240° to 300°   (g) 300° to 360°

Figure 3.6 ON diode chart and switch-equivalent circuits of the three-phase full-wave rectifier.

circuits of the three-phase full-wave rectifier are shown in Fig. 3.6(b) through (g). These schematics may be constructed from the chart shown in Fig. 3.6(a). From these equivalent-switch schematics it may easily be determined what the output voltage is at various 60° intervals. This output voltage may be taken from the graph in Fig. 3.7(a) and redrawn in Fig. 3.7(b) to form the output waveform.

For example, Fig. 3.6(b) shows that the voltage applied across the load

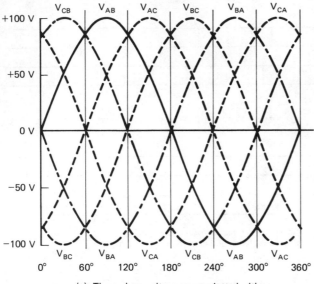

(a) Three-phase voltage source plotted with the three inverted voltages

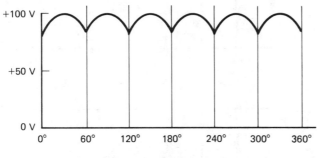

(b) dc output voltage waveform across R in Fig. 3.4 (a)

Figure 3.7 Analysis of the voltage output of the three-phase full-wave rectifiers.

is from the source terminals C to B. That is, from $0°$ to $60°$ the output is V_{CB}. Hence, that portion ($0°$ to $60°$) of V_{CB} from Fig. 3.7(a) is redrawn in Fig. 3.7(b). Figure 3.6(c) shows the output to be V_{AB} from $60°$ to $120°$. Figure 3.6(d) shows the output from $120°$ to $180°$ to be V_{AC}, from $180°$ to $240°$ to be V_{BC}, from $240°$ to $300°$ to be V_{BA}, and from $300°$ to $360°$ to be V_{CA}.

3.4 THREE-PHASE DUAL-WAVE RECTIFIERS

Figure 3.8 shows the schematic of the three-phase, dual-wave rectifier. The transformer connections are simplified to omit extra crossing wires. Each primary winding has two secondary windings as shown on the left. The three-turn windings are connected in the wye configuration with the diodes as shown by the letter codes. The four-turn windings are letter coded to the delta circuit on the right.

The output frequency of the dual-wave is twelve times its input. It is made up of two three-phase full-wave rectifiers like the ones shown in Figs. 3.2 and 3.3. The primary wye-phase voltages and the delta-phase voltages are in phase. The delta-line voltages are its phase voltages, but the wye-line voltages are thirty degrees from the wye-phase voltages. Since there are six (including negatives) wye voltages and six delta voltages, there are twelve voltages all 30° apart. This yields a very smooth rectified output.

Figure 3.9 shows more specifically how the twelve-phase output is developed vectorially. The primary-transformer voltages are shown in Fig. 3.9(a) as a vector reference. Figure 3.9(b) shows the three delta voltages in solid lines and their negatives in dotted. Figure 3.9(c) shows the wye-phase voltages in solid lines. These are in phase with the primary vectors and the delta vectors. The wye negative-voltage vectors are shown in dotted lines. The wye-line voltages are shown as resultants with short and long broken lines. The resultants are the vector sums of the phases and the negative vectors.

Note that the wye-phase voltages are smaller than the delta (see Fig. 3.9(b) and (c)]. The delta secondary windings have the $\sqrt{3}$ times more turns than the wye. Then, when wye voltages are added, their resultants are equal to the delta voltages but 30° from them [see Fig. 3.9(c)].

In the rectifier circuit shown in Fig. 3.8, the delta and wye add vectorially. This final vector addition is shown in Fig. 3.9(d) with the resultant voltages shown in solid lines.

3.4.1 Analysis of the Three-Phase Dual-Wave Rectifier

The analysis of the dual-wave follows the same procedure as the three-phase full-wave rectifier (see Sec. 3.3.1). First, the terminal voltages of each of the two full-wave rectifiers are graphed as in Fig. 3.10 (all three terminal voltages. Then, it can be determined which diodes are on and when. Two diodes in the wye and two diodes in the delta, totalling four diodes, are on at all times. The chart in Fig. 3.10(c) summarizes the ON diodes. From this chart the twelve switch-equivalent circuits are drawn as in Fig. 3.11.

The twelve phases are shown in Fig. 3.12(a). Using the switch-equivalent circuits it can be determined which diodes are on during each 30° interval (see Sec. 3.3.1) and which sources are across R. These voltages are then redrawn and connected as shown in Fig. 3.12(b). This graph is the dc output-voltage waveform of the three-phase dual-wave rectifier.

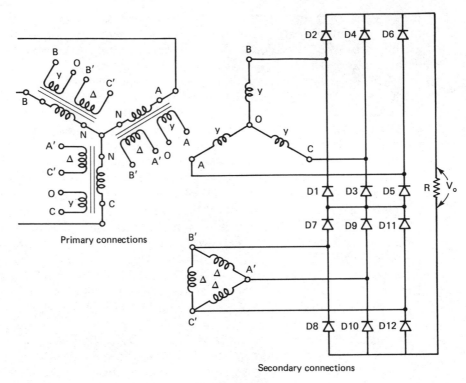

Primary connections

Secondary connections

(a) Schematic diagram

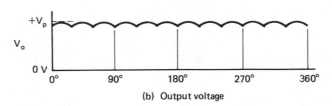

(b) Output voltage

Figure 3.8 The three-phase dual-wave rectifier.

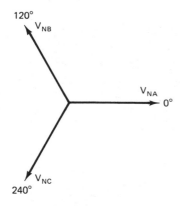

(a) Transformer primary-voltage vectors

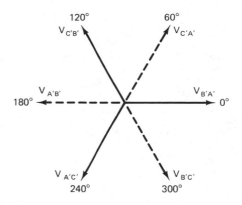

(b) Transformer secondary-voltage vectors on the delta connection

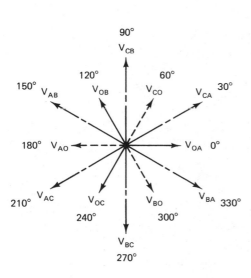

(c) Transformer secondary-voltage vectors on the wye connection showing resultant three-wire vectors

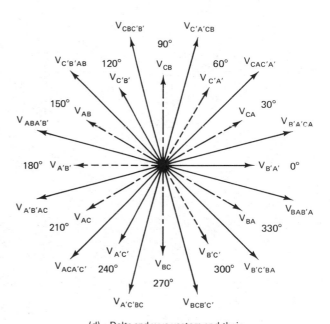

(d) Delta and wye vectors and their twelve resultant vectors

Figure 3.9 The addition of the delta and the wye voltage vectors showing the twelve resultant voltages which drive the three-phase dual-wave rectifier.

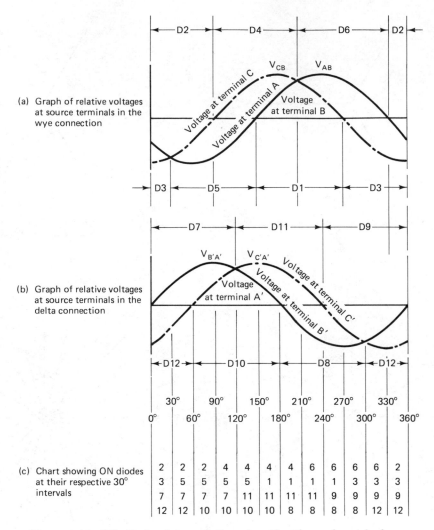

(a) Graph of relative voltages at source terminals in the wye connection

(b) Graph of relative voltages at source terminals in the delta connection

(c) Chart showing ON diodes at their respective 30° intervals

2	2	2	4	4	4	4	6	6	6	6	2
3	5	5	5	5	1	1	1	1	3	3	3
7	7	7	7	11	11	11	11	9	9	9	9
12	12	10	10	10	10	8	8	8	8	12	12

Figure 3.10 ON diode determination for the three-phase dual-wave rectifier.

34

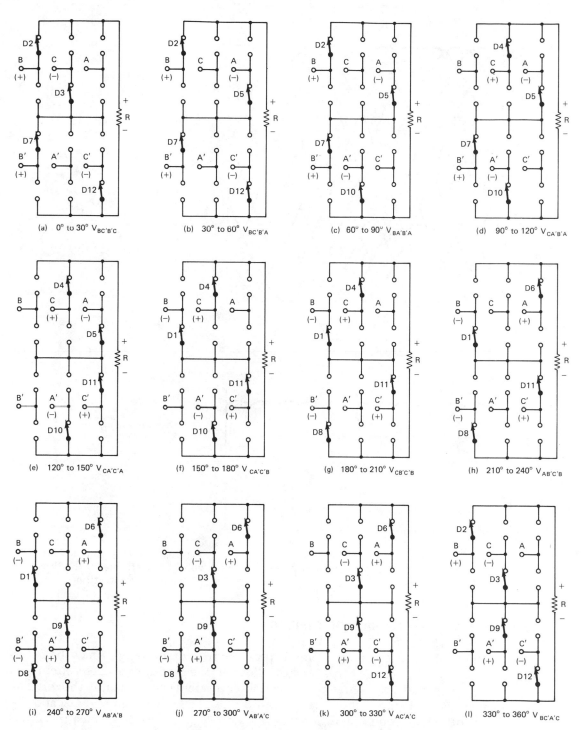

Figure 3.11 The diode switch-equivalent circuits of the three-phase dual-wave rectifier showing current paths for one complete cycle of the input voltage.

35

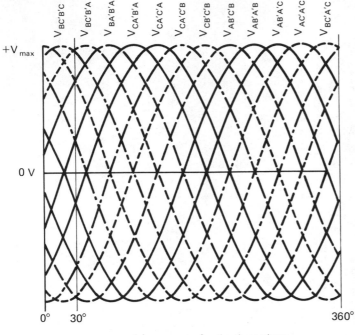

(a) ac source for the three-phase
dual-wave rectifier

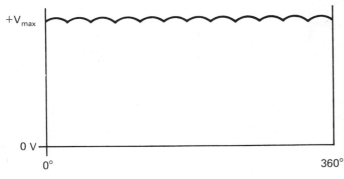

(b) dc output voltage for the three-phase
dual-wave rectifier

Figure 3.12 The input and output
voltage waveforms for the three-
phase dual-wave rectifier.

3.5 PIVC OF THE THREE-PHASE RECTIFIERS

The peak-inverse voltage a circuit presents to a diode (PIVC) may be determined by the voltage across any off diode in the switch-equivalent circuit.

The half-wave three-phase switch-equivalent circuit is not shown. The
analysis of Fig. 3.1 shows that from $30°$ to $150°$ D_1 is on and D_2 and D_3
are off. This connects V_{AB} across D_2 and V_{AC} across D_3. Thus, the PIVC
of the half wave is $\sqrt{3}$ peak-phase voltage or just the peak-line voltage.

Figure 3.6(b) is the switch equivalent of the full-wave three-phase rectifier. It shows V_{CA} across D_2, V_{CB} across D_4 and D_5, and V_{AB} across D_1. Again, the PIVC is the peak-line voltage or delta-source voltage.

Figure 3.11(a) shows V_{BC} across D_1 and $V_{B'A'}$ across D_9. These peak voltages will be the greatest any diode will encounter in the dual-wave three-phase rectifier. V_{BC} equals $V_{B'A'}$.

Thus for any three-phase rectifier the PIVC is the peak-line or delta voltage, or $\sqrt{3}$ times the peak-wye voltage source.

PROBLEMS

3.1. List three advantages of three-phase rectifiers.

3.2. If V_{BN}, V_{AN}, and V_{CN} in Fig. 3.1 are 100 V each, what is the value of V_{AB}?

3.3. List the three types of three-phase connections.

3.4. What two diodes are on from 90° to 120° in Fig. 3.2?

3.5. What two diodes are on from 240° to 300° in Fig. 3.3?

3.6. Sketch the three switch-equivalent circuits of the circuit shown in Fig. 3.1(a).

3.7. What odd-numbered diode is never on when D_4 in Fig. 3.4(a) is on?

3.8. Which terminal has the most positive voltage in Fig. 3.3 at 150°? (*HINT*: see Fig. 3.5.)

3.9. What is the phase shift, in degrees, from V_{AB} to V_{BA}?

3.10. Explain why secondary winding OA has less turns than winding $B'A'$ in Fig. 3.8(a).

3.11. List all of the source terminals and components in proper sequence showing the path of current through the circuit shown in Fig. 3.11(c). Begin at B.

3.12. How many degrees is $V_{B'A}$ from $V_{CA'}$ in Fig. 3.8(a)? Consider 0 to C' and A' to 0 connected in order to answer this question.

3.13. List the six secondary voltages which are in phase with the three primary voltages (V_{NA}, V_{NB}, and V_{NC}) shown in Fig. 3.8(a).

3.14. Find the exact turns ratio of the delta-secondary-to-wye-secondary in Fig. 3.8(a) so that the magnitude of the wye V_{BC} equals the delta $V_{B'C'}$.

3.15. List the three three-phase rectifiers and their respective frequency outputs compared to one sine-wave cycle of phase A.

3.16. Why are the graphs of the three-phase voltages redrawn with respect to phase B in Fig. 3.5?

3.17. Explain why D_2 and D_4 are off when D_6 is on in the dual-phase circuit shown in Fig. 3.8(a).

Special Diodes

4

This chapter covers five special diodes which are commonly used in power electronics. The *Zener* is used to regulate dc supply voltages. It is commonly used in trigger oscillators [see Fig. 9.3(a)]. The *thyrector* arrests transient spikes, regulating ac source voltages. The two types of *diacs* and the *Shockley* diode are the active devices in some trigger oscillators used to turn on high-power devices at desired periods.

4.1 REGULATORY DIODES

4.1.1 The Zener Diode

The Zener diode has an accurate and stable reverse-breakdown voltage. This makes it useful as a voltage standard and/or a voltage regulator. Its rated-breakdown voltage [see Fig. 4.1(b)] is usually within 5% of its nominal voltage. For example, a 9 volt type of Zener may have a breakdown rating from about 8.95 V or 9.05 V. However, if one particular Zener breaks down at 9.04 V, it will do it consistently within less than 1% tolerance. Once a particular Zener is calibrated, it is a very accurate voltage standard.

Figure 4.1(a) shows the schematic of a Zener diode. Figure 4.1(b) shows its characteristic curve and breakdown voltage. Figure 4.1(c) shows a Zener being used as a voltage regulator. The Zener must be (1) reverse-biased, with (2) a voltage greater than its Zener voltage (breakdown voltage), and (3) have a series resistance to limit current below its rated value.

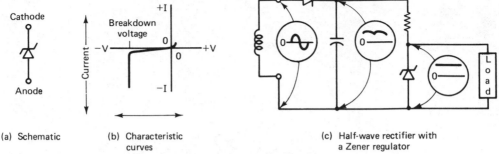

| (a) Schematic | (b) Characteristic curves | (c) Half-wave rectifier with a Zener regulator |

Figure 4.1 The Zener diode schematic, volt-ampere characteristic curve, and in an application as a regulator.

4.1.2 Zener Diode Circuit Analysis

Replacing a faulty Zener diode should be done by a technician who is able to analyze the circuit. Example 4-1 covers three pitfalls of faulty installation and simplifies Zener circuits in general.

Example 4-1

Find the current through the Zener diode in Fig. 4.2 and check the circuit for proper operation when: the Zener voltage (V_Z) is 6 V, maximum rated current (I_Z) is 50 mA, source voltage (E) is 20 V, R_1 is 330 Ω, and the load resistance (R_L) is 220 Ω.

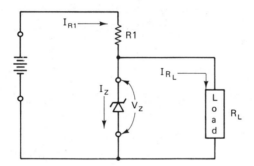

Figure 4.2 Zener diode as a voltage standard.

Solution.

1. Check the bias polarity of the Zener. Its anode is connected to the negative end of the source, reverse biasing the Zener. Bias polarity is okay.

2. Find the open-circuit voltage across the Zener and see if it is larger than V_Z. With the Zener removed, the voltage across its terminal connections (V_{ZOC}) is

$$V_{ZOC} = E \frac{R_L}{R_1 + R_L} = 20 \text{ V} \frac{220}{330 + 220} = 8.0 \text{ V}$$

Hence, the value of reverse voltage is large enough to enable the Zener to conduct in its reverse-breakdown region.

40

3. With the Zener connected, assume V_Z across it and find its current.

(a) First, find the current through R_1:

$$I_{R_1} = \frac{E - V_Z}{R_1} = \frac{20\text{ V} - 6\text{ V}}{330} = \frac{14\text{ V}}{330} = 42.42\text{ mA}$$

(b) Find the current through the load (R_L):

$$I_{R_L} = \frac{V_Z}{R_L} = \frac{6\text{ V}}{220} = 27.27\text{ mA}$$

(c) Find I_Z:

$$I_Z = I_{R_1} - I_{R_L} = 42.42\text{ mA} - 27.27\text{ mA} = 15.15\text{ mA}$$

I_Z is less than $I_{Z(\text{max})}$, thus the circuit is correct.

4.1.3 The Thyrector

Figure 4.3 shows the schematic of the thyrector. It operates like two Zeners connected back-to-back in series. It is used to suppress voltage surges and transients. It acts the same toward rated voltages of either polarity. One diode is always forward-biased, acting like an ON diode. Then, the second diode acts like a Zener. The suppressor effect is to conduct for all overvoltages and drop them to a rated value.

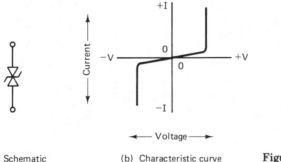

(a) Schematic   (b) Characteristic curve   **Figure 4.3** The thyrector.

4.2 TRIGGER DIODES

4.2.1 The Diac

The diac operates like two diodes connected back-to-back in series. Unlike the thyrector these diodes are not like Zeners. Zener diodes maintain the same breakdown voltage after breakdown. Some common earlier diodes have breakdown characteristics as shown in Fig. 4.4. After these diodes break down, higher currents flow causing higher temperatures which lower breakdown voltages.

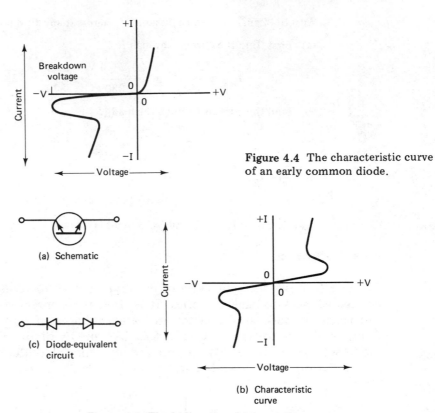

Figure 4.4 The characteristic curve of an early common diode.

(a) Schematic

(c) Diode-equivalent circuit

(b) Characteristic curve

Figure 4.5 The bidirectional-trigger diac.

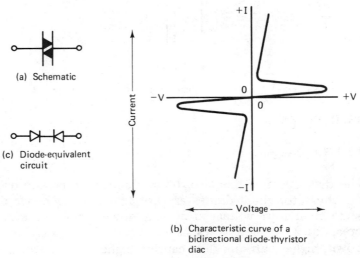

(a) Schematic

(c) Diode-equivalent circuit

(b) Characteristic curve of a bidirectional diode-thyristor diac

Figure 4.6 The bidirectional diode-thyristor diac.

The two types of diacs are shown in Figs. 4.5 and 4.6. Figure 4.5 shows the schematic, characteristic curve, and diode-equivalent circuit of the bidirectional-trigger diac. Figure 4.6 shows the schematic, characteristic curve, and diode-equivalent circuit of the bidirectional diode-thyristor diac. The thyristor curve is more pronounced. Its ON voltage is lower than the trigger diac. Hence, it dissipates less power for the same current. This enables it to be used with heavier loads.

Both types of diacs are used to trigger larger thyristors such as SCRs and triacs (see Sec. 14.4).

4.2.2 The Shockley Diode

The Shockley diode is also a trigger device. It is used to turn on a circuit when its forward voltage is exceeded. Figure 4.7 shows the schematic, the *p-n* diagram, transistor equivalent, and characteristic curve.

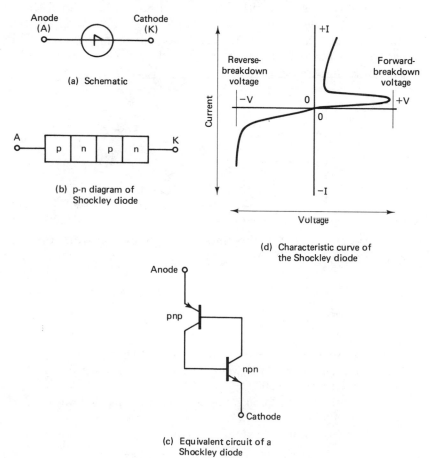

Figure 4.7 The Shockley diode.

The Shockley diode acts like a normal diode when it is reverse-biased. When it is forward-biased it blocks current until its forward-breakover voltage is exceeded. Then, it conducts readily [see Fig. 4.7(d)]. A common breakover voltage is 60 V.

The transistor-equivalent circuit [see Fig. 4.7(c)] shows that *pnp* transistor base is the *npn* collector and the *npn* base is the *pnp* collector. As forward voltage nears breakover, the leakage current through the device reaches the turn-on value for both transistors. Once both transistors turn on the device latches. That is, the *pnp* transistor turns on furnishing the *npn* with base current, and the *npn* turns on furnishing the *pnp* with base current. Thus, the two transistors hold each other on until anode-to-cathode current goes to zero or at least very close to zero.

Figure 4.8 shows a simple application for the Shockley diode. When S_1 is closed, the supply voltage (E) furnishes current through R charging up C. When the voltage across C gets to the Shockley turn-on voltage, the diode turns on. This shorts out C, quickly discharging it until the charge on C is too low to keep the Shockley diode in conduction. The Shockley diode turns off allowing the capacitor C to repeat its charging. This sequence of events repeats itself indefinitely. The dc source E is changed to a sawtooth output as shown in Fig. 4.8.

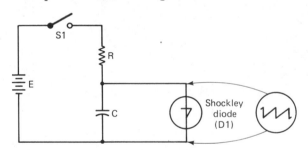

Figure 4.8 The Shockley diode used in an oscillator.

Another use for the Shockley diode is to turn on (trigger) a larger thyristor such as an SCR.

PROBLEMS

4.1. List two applications of the Zener diode.

4.2. Find the current through the Zener in Fig. 4.2 when: E = 28 V, V_Z = 18 V, R_1 = 200 Ω, and R_L = 600 Ω.

4.3. Find the current through the Zener diode in Fig. 4.2 when: E = 20 V, V_Z = 16 V, R_1 = 400 Ω, and R_L = 2 kΩ.

4.4. Sketch the schematic of the thyrector.

4.5. Sketch the schematic of the bidirectional-trigger diac.

4.6. Sketch what you feel the *p-n* diagram of the bidirectional-trigger diac would be.

4.7. Sketch the diode-equivalent circuit of a diac.

4.8. Sketch the schematic of the Shockley diode.

4.9. What device(s) does the thyrector act like?

4.10. What device(s) does a Shockley diode act like?

4.11. Describe how a Shockley diode is turned on.

Transistors

5

The transistor has two basic types of applications: switching and amplifying. The switching applications are important in power electronics.

Diodes are switches but have only two terminals. They only respond to (switch) the voltage across them. Transistors have three terminals. Two terminals act like switch contacts. The third terminal is used to actuate the other two. Thus, the control circuit may be independent of the circuit being controlled. The switching transistor is like a *relay*, which has one terminal common to both its coil and one contact [see Fig. 5.1(a)].

5.1 BIPOLAR TRANSISTORS

The type of transistor shown in Figs. 5.1 and 5.2 is called the *bipolar transistor*. The name bipolar is usually omitted as the other types of transistors have special names and are less common than the bipolar. The name transistor implies the bipolar transistor, unless otherwise noted.

The transistor has three elements called the *base*, *emitter*, and *collector*. The base is shown as a line or bar at right angles to the base lead. The emitter always has an arrowhead on it. If the arrowhead points toward the base it is a *pnp* transistor [see Fig. 5.2(c)]. If it points away from the base it is an *npn* transistor. The collector is always shown at an angle to the base, just as the emitter is, but the collector has no arrowhead.

Figure 5.1(b) is the schematic of an *npn* transistor. A small current through the base to the emitter turns on the collector-to-emitter path. This

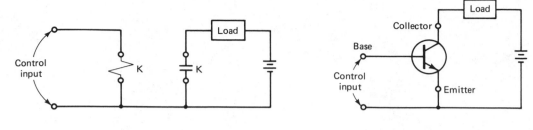

(a) Relay analogy of a switching transistor   (b) Switching transistor

Figure 5.1 A switching transistor used in place of a relay.

path is capable of carrying many times more current than the base-emitter junction.

Figure 5.2 shows schematics of both *pnp* and *npn* transistors along with their respective diode analogies. Two diodes connected as in Fig. 5.2(b) or in 5.2(d) could not replace a transistor. These analogies are used to show how to test a transistor with an ohmmeter and/or describe how a transistor is biased.

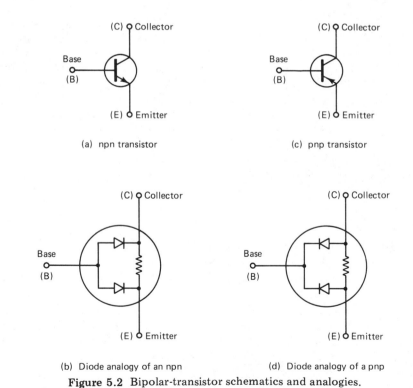

(a) npn transistor   (c) pnp transistor

(b) Diode analogy of an npn   (d) Diode analogy of a pnp

Figure 5.2 Bipolar-transistor schematics and analogies.

A transistor may be tested with an ohmmeter by testing the base-emitter diode and then the base-collector diode, as described in Sec. 2.1. Finally, the resistance between the emitter and collector (R_{EC}) may be tested. This resistance is much greater than the forward resistance of either diode. In silicon transistors R_{EC} may read infinity on some ohmmeters. Faulty power transistors often appear shorted from collector to emitter, even when both diodes test OK.

The collector-to-emitter path of a transistor differs from relay contacts in that they are polarized. In an *npn* the collector must be more positive than the emitter and vice versa in a *pnp*.

A more complete biasing rule is that *the base-emitter junction must be forward-biased, while the base-collector must be reverse-biased for the transistor to be on.* This rule holds for both *npn* and *pnp* transistors [see Fig. 5.2(b) and (d)].

The base-emitter junction acts like a diode (see the diode characteristic curve shown in Fig. 2.2. For voltages over 0.7, in silicon, a current-limiting resistor should be used in series with the base-emitter junction.

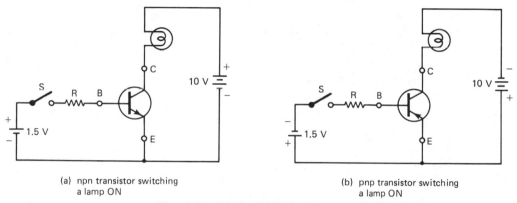

(a) npn transistor switching
a lamp ON

(b) pnp transistor switching
a lamp ON

Figure 5.3 Biasing a bipolar transistor ON.

Figure 5.3 shows how an *npn* and a *pnp* transistor is biased on. In either case the base-emitter junction is forward-biased when switch S is closed. Closing S turns the lamp on. In both cases the base-emitter junction is forward-biased and the base-collector junction is reverse-biased. An ON transistor has a slightly larger drop across its collector-to-emitter terminals. With S closed, the base B is at about 0.7 V with respect to its emitter, while its collector is about the same. This reverse-biases the collector-base junction as it takes a positive 0.7 V from collector to base to forward-bias it.

The collector and the emitter in a bipolar transistor are not reversible. The transistor's characteristics and ratings change significantly when these two terminals are reversed.

5.2 FIELD EFFECT TRANSISTORS (FETs)

There are two basic types of *field-effect transistors* (FETs). These are the *junction FET* (JFET) and the *insulated gate FET* (IGFET). Insulated gate FETs have higher input impedances and some types may respond to positive gate signals.

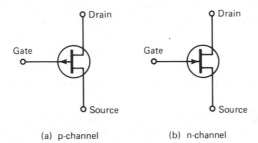

(a) p-channel   (b) n-channel

Figure 5.4 Schematic diagrams of the junction field-effect transistor (JFET).

The two schematics of the JFET are shown in Fig. 5.4. Figure 5.4.(a) shows a *p*-channel JFET. The three electrodes of an FET are the *gate*, *source*, and *drain*. With zero volts on the gate, the path from source to drain has low resistance and the JFET is on. When a minus, with respect to the source voltage is on the gate, the source-to-drain path has a higher resistance and may be turned completely off.

The FET source and drain are interchangeable. That is, the source and drain are not designated until they are connected in the circuit. The source is always at the potential nearest the gate. The drain is connected to the load.

5.3 UNIJUNCTION TRANSISTORS (UJTS)

The *unijunction transistor* (UJT) is used as an oscillator and sometimes used as a current or voltage sensor. The UJT oscillator is used to turn on larger devices such as an SCR (see Chapter 6).

Figure 5.5(a) shows the schematic diagram of the UJT. The UJT has three electrodes: the emitter (E), base one (B_1), and base two (B_2). Its control terminals (E and B_1) are the same as its controlled path. Terminal B_2 is used for biasing and temperature compensation.

There is a nominal resistance between B_2 and B_1. This is called the *interbase resistance* (R_B). It is made up of two resistances: R_{B_1} and R_{B_2} [see Fig. 5.5(b)]. The ratio of R_{B_1} to R_B is called the eta (η) ratio where

$$\eta = \frac{R_{B_1}}{R_B} = \frac{R_{B_1}}{R_{B_1} + R_{B_2}} \tag{5.1}$$

The eta ratio is 0.6 in most UJTs.

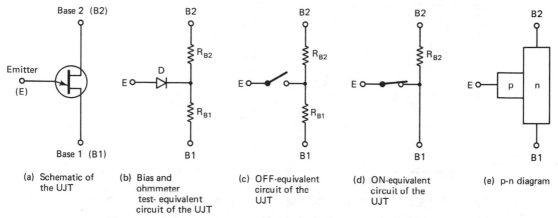

| (a) Schematic of the UJT | (b) Bias and ohmmeter test- equivalent circuit of the UJT | (c) OFF-equivalent circuit of the UJT | (d) ON-equivalent circuit of the UJT | (e) p-n diagram |

Figure 5.5 The schematic diagram and equivalent circuits of the UJT.

Figure 5.5(b) is an equivalent circuit of a UJT. This circuit (a diode and two resistors) cannot be used to replace a UJT. It can be used to describe how a UJT may be biased and how the UJT may be tested with an ohmmeter.

There are five ohmmeter resistances to check when testing a UJT:

1. B_1 to B_2 (R_B)
2. E to B_1 with D forward-biased (R_{B1_F})
3. E to B_1 with D reverse-biased (R_{B1_R})
4. E to B_2 with D forward-biased (R_{B2_F})
5. E to B_2 with D reverse-biased (R_{B2_R})

R_{B1_R} and R_{B2_R} should be infinite or extremely large compared to the other three resistances. R_{BB} is the third largest resistance (approximately 10,000 Ω). R_{B1_F} and R_{B2_F} are the smallest resistances. The value of these smaller resistances varies widely as the resistance of a forward-biased diode depends on the current through it. With equal currents R_{B1_F} should be about 20% larger than R_{B2_F}.

Figure 5.5(c) shows the UJT when it is OFF, and Fig. 5.5(d) is the UJT-equivalent ON circuit. Figure 5.5(e) is the UJT p-n diagram.

Figure 5.6 shows how the UJT is usually biased. A 10-V source is connected across B_1 to B_2. The resistance R_{B_1} and R_{B_2} act as a voltage divider. This places 0.6 (eta ratio) of the supply voltage across R_{B_2}. Thus, the voltage at the eta point is 6 V with respect to ground. This places +6 V on the cathode of the germanium diode. It takes about 0.2V to forward-bias and turn it on. Therefore, the anode (emitter) must be greater than +6.2 V. The potentiometer wiper, shown in Fig. 5.6, may be raised to yield 6.2 V. This

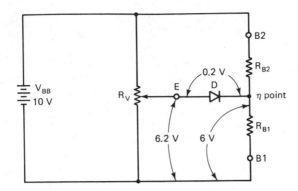

Figure 5.6 Biasing the UJT.

turns the UJT on. After the UJT is turned on, it acts like a forward-biased diode from E to B_2 [see Fig. 5.5(d)].

Figure 5.7 shows the UJT oscillator circuit. This particular circuit is ideal for triggering an SCR. It is a very temperature- and source-voltage-stable circuit and has an output suited for SCR firing.

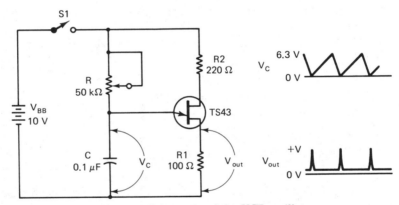

Figure 5.7 Schematic of the UJT oscillator.

When the switch S is turned on, C started to charge through R. The voltage-divider circuit of R_1, R_2, and the UJTs R_{B_1} and R_{B_2} sets up 6 V at the UJT's eta point. When the charge on C gets to about 6.3 V the UJT turns on. This places a low resistance across C and quickly discharges it. This turns the UJT off, C starts to charge again, and the process repeats. The voltage across C (V_C) is a sawtooth caused by relatively slow charge and rapid discharge (see V_C in Fig. 5.7). During the rapid discharge, a high current through R_1 causes a voltage-spike output (see V_{out}) in Fig. 5.7.

The time it takes to turn on the UJT is about one time constant of R and C (see Fig. 5.7). Thus, where $\eta = 0.6$

$$T = RC \tag{5.2}$$

The emitter trigger voltage is about 63% of V_{BB}. Frequency (f) is the reciprocal of time. Thus, f is approximately

$$f = \frac{1}{RC} \tag{5.3}$$

PROBLEMS

5.1. List the three electrodes of the bipolar transistor.

5.2. Describe how the base-emitter junction and base-collector junction must be biased to turn on the collector to emitter.

5.3. What electrode is at the most negative voltage in an ON *pnp* transistor?

5.4. What two bipolar-transistor electrodes act like the coil terminals of a relay?

5.5. What two bipolar-transistor electrodes act like the contact of a relay?

5.6. What two FET electrodes act like the coil terminals of a relay?

5.7. What two FET electrodes act like the contacts of a relay?

5.8. What two UJT terminals are the control terminals?

5.9. What two UJT electrodes are the controlled terminals?

5.10. What is B_2 used for on a UJT?

5.11. List two applications of a UJT.

5.12. What is the difference between the schematics of the FET and the UJT?

5.13. Which transistor has two interchangeable electrodes?

5.14. What is the frequency of the UJT oscillator shown in Fig. 5.7?

Thyristor Devices

6

Thyristors are *pnpn* devices used as electronic switches. Their main advantage is the control of large amounts of power using a very small input power. The Schockley diode (see Sec. 4.2.2) is a thyristor. This chapter serves as an introduction to the thyristor family. The SCR is by far the most popular thyristor and is emphasized throughout the following chapters. SCRs are used to control electric furnaces, heaters, lighting, X-ray power, and electric motor speed, torque, and acceleration. Its operating mechanisms are used to explain how other thyristors work.

6.1 SCR (SILICON CONTROLLED RECTIFIER)

6.1.1 Introduction to the SCR

The *silicon controlled rectifier* (SCR) is the most popular electrical power controller. One single device can control up to a million watts of electrical power. Arrays of SCRs have been used to control loads greater than 100 MW.

SCRs have been used to replace rheostats, magnetic amplifiers, gas-filled tubes, and autotransformers; many types of motor generator sets such as inverters, converters, and rectifiers; relays, contactors, circuit breakers, fuses, and flashers; logic devices such as counters, registers, and memory devices.

Figure 6.1 shows the equivalent circuits and schematic diagram of the SCR. It has three electrodes: gate (G), anode (A), and cathode (K). A positive voltage on its gate with respect to its cathode, turns on (triggers) the

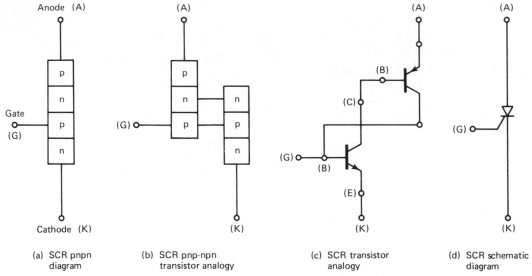

Figure 6.1 SCR-equivalent circuits and schematic diagram.

anode-to-cathode path when the anode is positive. However, the current through the anode-to-cathode path cannot be turned off using the gate. It is turned off by interrupting the anode current. The SCR acts like an electrical-latching relay.

The SCR operates like two transistors as shown in Fig. 6.1(b) and (c). A positive voltage on the gate forward-biases the base-emitter junction of the *npn* transistor, turning it on. This allows current through the *npn* collector (the *pnp* base). If the SCR anode is positive, the *pnp* emitter-base junction is forward-biased, turning it on. After the *pnp* transistor is turned on, it in turn supplies the *npn* with base current. With gate voltage and current removed, the SCR is still in conduction from anode to cathode. The *npn* supplies the *pnp* with base current and the *pnp* supplies the *npn* with base current. The SCR remains on until its principal current is interrupted.

The SCR principal current (current from anode to cathode) may be interrupted three ways:

1. shorting the SCR from anode to cathode
2. opening the external path from its anode supply voltage (V_{AA})
3. V_{AA} going negative, as it does, if it is an ac source

Figure 6.2 shows a family of SCR-characteristic curves. The only differences in the three curves are in the + V_{AK} region. When gate current is zero (I_G = 0) the SCR blocks just as a Shockley diode (see Sec. 4.2.2). At higher gate currents (I_G = 1 and I_G = 2) the SCR turns on (fires) at lower forward voltages. In the reverse direction it acts like a diode.

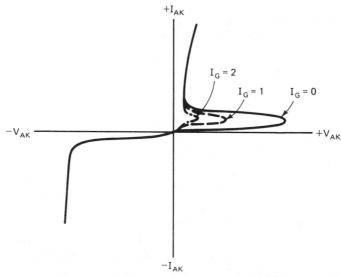

Figure 6.2 Characteristic curve of the SCR.

6.1.2 Testing SCRs for Proper Latching

Small SCRs may be quickly tested with an ohmmeter or for larger SCRs the simple circuit in Fig. 6.3 may be used. These tests consist of forward-biasing the SCR from anode to cathode, supplying the SCR with a short duration gate-trigger voltage, and observing if the SCR latches on (stays on) after the gate-trigger source is disconnected.

An ohmmeter may be used to test SCRs which have low holding cur-

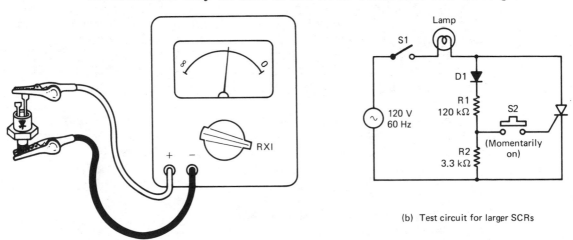

(a) Testing low-current SCRs with an ohmmeter

(b) Test circuit for larger SCRs

Figure 6.3 Testing SCRs for proper latching.

rents (see Sec. 7.2.5). Generally, most SCRs with a maximum principal current rating of less than 20 A may be tested with a Simpson 260 VOM on its RX1 range. The ohmmeter is convenient because it serves three functions:

1. to forward-bias the SCR from anode to cathode
2. supply the SCR gate-trigger voltage
3. to indicate SCR conduction

Figure 6.3(a) shows an SCR being tested with an ohmmeter. First, select the lowest RX range on the ohmmeter. This provides the largest current from the ohmmeter. Connect the negative-ohmmeter lead to the SCR cathode. Connect the positive-ohmmeter lead to the SCR anode. Short the SCR anode, momentarily, to the gate. The SCR should conduct and remain on after the short from the anode to the gate is removed. The anode and cathode must remain connected to the ohmmeter during the test.

If the SCR conducts with no gate-trigger voltage, it is shorted from anode to cathode. If it does not conduct, even with a gate-trigger voltage, it is open. If it conducts during gate trigger, but turns off when the trigger is removed, its condition is questionable. Holding current supplied by the ohmmeter may not be large enough, if it conducts only during triggering. Then, it should be tested with a circuit such as the one shown in Fig. 6.3(b). If it is open or shorted it must be replaced if it cannot be repaired.

If the SCR does not conduct at all, there is a possibility that the trigger-circuit current is too low. The trigger voltage in Fig. 6.3(b) may be increased by decreasing R_1.

For the test circuit shown in Fig. 6.3(b) the components used must have special ratings (see Chapter 7). The diode and SCR must have at least a PIV of 200 V, the lamp should have a hot resistance low enough to draw the SCR holding current (I_H) and less than the SCR I_{AV} maximum rating.

In order to test an SCR in the circuit in Fig. 6.3(b), connect the circuit as shown. Then, turn on S_1. Next, depress S_2 momentarily. The lamp should light and remain on until S_1 is turned off. If this happens the SCR is working properly.

SCRs may be tested more thoroughly with a transistor curve tracer. This may be necessary if the exact values of certain voltages or currents must be tested. But, for a fast check of SCRs, the tests shown in Fig. 6.3 work well.

6.2 PUT (PROGRAMMABLE UNIJUNCTION TRANSISTOR)

The *programmable unijunction transistor* (PUT) is actually a thyristor. It is a *pnpn* device such as the SCR. The major difference is that the PUT has its external-gate electrode connected to its *n*-type material nearest the anode.

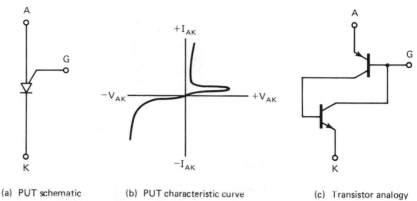

(a) PUT schematic (b) PUT characteristic curve (c) Transistor analogy

Figure 6.4 Programmable unijunction transistor (PUT).

The PUT schematic, PUT characteristic curve, and transistor analogy circuit are shown in Fig. 6.4.

A second major difference in the PUT and the SCR lies in their applications. The PUT is used in time delay, logic, and SCR trigger circuits. The largest PUT can only handle about 200 V and less than 1 A.

6.3 SPECIAL SCRs

6.3.1 The Amplifying-Gate SCR

The *amplifying-gate SCR* is shown in Fig. 6.5. The amplifying-gate SCR has the advantage of being triggered by very small gate-trigger currents, less than 50 μA, while the devices available have fairly impressive capabilities (1000 A and up to 1200 V). Its schematic is an SCR with a diode pointing toward its gate. Its characteristic curves are identical to the SCR [see Fig. 6.5(b)]. Its equivalent circuit is a small SCR driving a larger SCR's gate [see Fig. 6.5(c)].

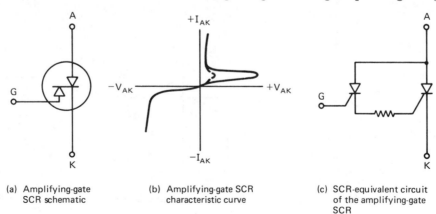

(a) Amplifying-gate SCR schematic (b) Amplifying-gate SCR characteristic curve (c) SCR-equivalent circuit of the amplifying-gate SCR

Figure 6.5 Amplifying-gate SCR.

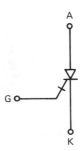

Figure 6.6 Gate-controlled switch (GCS) or gate turn-off SCR (GTO).

6.3.2 The Gate Turn-Off SCR (GTO)

The schematic of the *gate turn-off SCR* (GTO) or sometimes known as the *gate-controlled switch* (GCS) is shown in Fig. 6.6. Its capabilities are less than 20 A and less than 600 V. It is only mentioned here as it may be found in some existing equipment. It has been largely replaced by high-power silicon transistors. The GTO is basically an SCR which can be turned off by a gate current. However, the gate current required to turn it off is nearly one-fifth as large as its principal current.

6.4 SUS (SILICON UNILATERAL SWITCH)

The *silicon unilateral switch* (SUS) is strictly a trigger, timing, or logic device. Its capabilities are less than 0.5 A and less than 20 V. The SUS is basically a PUT with a precisely controlled gate-trigger voltage. SCR trigger voltages and/or currents vary widely with changes in ambient temperature. The SUS triggers within a half of a percent of its rated value regardless of nominal temperature changes.

Figure 6.7 shows the schematic diagram, characteristic curve, and equivalent circuit of the SUS.

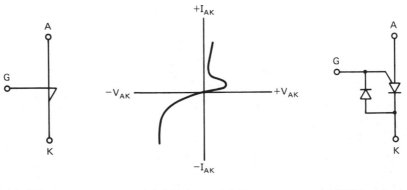

(a) SUS schematic   (b) SUS characteristic curve   (c) PUT Zener-equivalent circuit of the SUS

Figure 6.7 Silicon unilateral switch (SUS).

6.5 SBS (SILICON BILATERAL SWITCH)

Figure 6.8 shows the schematic, characteristic curve, and equivalent circuit of the *silicon bilateral switch* (SBS). Its capabilities and applications are like the SUS. It has one important advantage over the SUS in that it can be turned on in either direction. Thus, its name "bilateral." It may be used to trigger on an ac voltage. Its equivalent circuit is two SUS's back-to-back in parallel and common gates [see Fig. 6.8(c)].

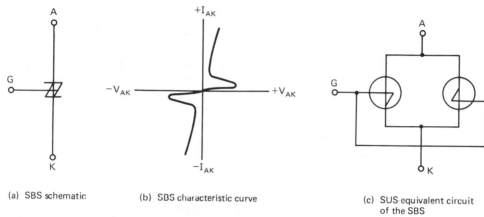

(a) SBS schematic   (b) SBS characteristic curve   (c) SUS equivalent circuit of the SBS

Figure 6.8 Silicon bilateral switch (SBS).

6.6 SCS (SILICON CONTROLLED SWITCH)

The *silicon controlled switch* (SCS) is a *tetrode* thyristor. That is, it has four electrodes. It has an anode gate (AG) like a PUT and a cathode gate (KG) like an SCR. A current of sufficient size on either gate will fire the SCS. A large reverse current through the anode gate may be used to turn the SCS

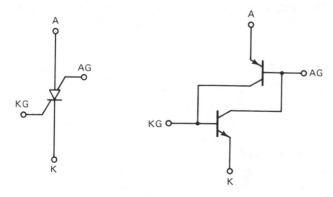

(a) SCS schematic   (b) Transistor-equivalent circuit of the SCS

Figure 6.9 Silicon controlled switch (SCS).

61

off. Its use is largely in voltage or current-sensing circuits as signals on either gate fire it. It operates like an OR gate. Its power capabilities are limited to timing, logic, and triggering applications. Figure 6.9 shows the schematic and equivalent circuit of the SCS. Its characteristic curve is like the SCR's.

6.7 TRIAC (BIDIRECTIONAL TRIODE THYRISTOR)

The *triac* is also a bidirectional thyristor. It has been used to replace the SCR in low-power ac applications such as small motor control or residential lamp dimmers. Its capabilities are less than 100 A and less than 1000 V. The triac's schematic is shown in Fig. 6.10(a). Its equivalent circuit shown in Fig. 6.10(c) describes how it is used. Actually, two SCRs cannot be connected this way unless their gates are independent. Chapter 14 is devoted to applications of the triac.

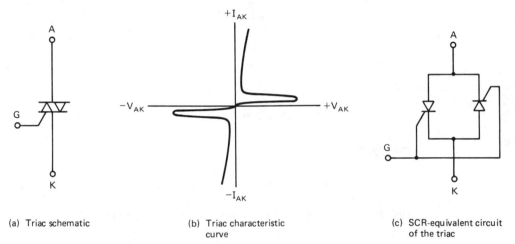

(a) Triac schematic   (b) Triac characteristic curve   (c) SCR-equivalent circuit of the triac

Figure 6.10 Bidirectional triode thyristor (triac).

6.8 LIGHT-ACTIVATED THYRISTORS

Four *light-activated thyristors* are shown in Fig. 6.11. The *light-activated switch* (LAS) is shown in Fig. 6.11(a). It has no external gate connection. Rather, light focused on its gate window latches it into conduction. It must be turned off as an SCR is turned off (see Sec. 6.1.2). Its characteristic curve is like the SCR.

The *light-activated SCR* (LASCR) is shown schematically in Fig. 6.11(b). It differs from the LAS in that it may be triggered with a light source or with its gate trigger. The gate may be biased with a voltage or current slightly less than its gate-trigger requirements, thereby varying the amount of light necessary to trigger it.

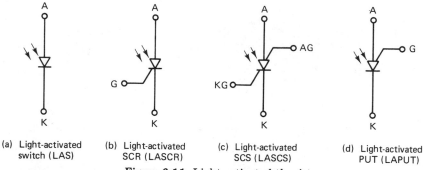

(a) Light-activated switch (LAS) (b) Light-activated SCR (LASCR) (c) Light-activated SCS (LASCS) (d) Light-activated PUT (LAPUT)

Figure 6.11 Light-activated thyristors.

Figure 6.11(c) is a schematic of the *light-activated SCS* (LASCS). It may be turned on or triggered three ways: via its anode gate, via its cathode gate, and/or via a light source through its light window. It may also be made to trigger via a combination of any two or all three of these trigger methods.

Figure 6.11(d) is the schematic of the *light-activated PUT* (LAPUT). It may be used as an ordinary PUT or be light activated, or use a combination of gate trigger and light source.

PROBLEMS

6.1. What is the most popular thyristor?

6.2. List six applications of the SCR.

6.3. Sketch the schematic diagram of the SCR.

6.4. Sketch the characteristic curve of the SCR with $I_G = 0$.

6.5. List three ways to turn off an SCR.

6.6. Sketch the schematic of a PUT.

6.7. Sketch the *pnpn* diagram of the PUT and label its three electrodes.

6.8. Sketch the schematic diagram of the amplifying-gate SCR.

6.9. What device has largely replaced the GTO?

6.10. What is the advantage of the SUS over the PUT?

6.11. List two bilateral thyristors.

6.12. List two tetrode thyristors.

6.13. Describe the difference between the LAS and the LASCR.

SCR Characteristics and Anode Ratings

7

7.1 SCR CHARACTERISTIC CURVES

Figure 7.1 shows the general SCR volt-ampere characteristics. There are actually three curves superimposed over each other. Each curve represents the anode-to-cathode voltage versus current at three different gate currents (I_G). All three curves are identical in all six regions except region # ③ , the forward-blocking region. This region distinguishes the SCR from the diode.

The $I_G = 0$ curve shows that the SCR can be turned on with no gate current. This is not a desirable characteristic in SCRs. It is a limiting one which must be considered. At a low value of gate current ($I_G = 1$) the SCR turns on at a lower value of forward-anode voltage. At a higher value of gate current ($I_G = 2$) the SCR fires at still a lower value of forward-anode voltage.

7.2 SCR ANODE RATINGS

7.2.1 SCR Rating Subscripts

Most of the important SCR ratings are given in terms of voltage (V) or current (I). Each of these ratings is accompanied by one, two, or three subscripts. The meaning of each letter subscript is given in the following list:

A usually stands for anode but may stand for ambient when used with a T for temperature

AV average

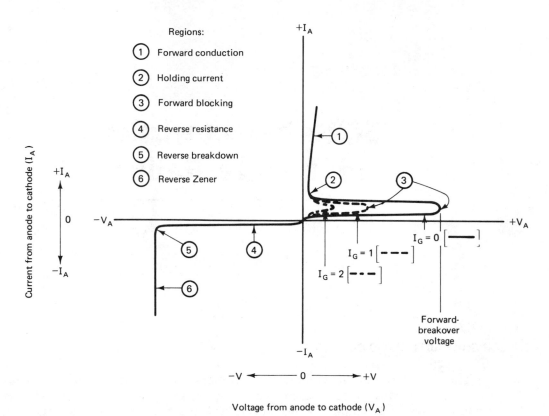

Figure 7.1 Characteristic curves of the SCR.

D represents the forward-blocking region when the SCR has no gate trigger

G gate

H holding

K cathode

L latching

M maximum

O represents open, for example, gate circuit open

R when used as the first subscript it stands for reverse, but when used as a second subscript it stands for repetitive

S when used as the first subscript it stands for shorted, but as a second subscript it stands for surge

T when used as the first subscript it stands for total or maximum, but as a second subscript it may stand for trigger

7.2.2 Maximum-Repetitive RMS-Current Rating

The most important SCR consideration is junction temperature (T_J). This is difficult to measure and control directly. However, the SCR voltages and currents which contribute to T_J are easier to measure and control. Probably the largest contributor to T_J is forward-repetitive rms-on-state current (I_{rms}).

SCR ratings are often given in terms of dc or average current. Average current is equal to rms in a steady-state dc circuit such as a dc battery. The average value of a positive (dc) pulse is easily measured with a dc ammeter. However, the rms or effective value of a pulse is much higher than its average value. The effective (rms) value determines heat dissipation.

There is great difficulty in measuring true rms current of a nonsinusoidal waveform (see Fig. 7.2). The rms value of this wave form may be approximated. Using a "worst case" approximation leaves a slight safety factor. The worst case approximation of the chopped-half sine wave is a rectangular wave with a height equal to the peak value and a width equal to the pulse duration of the nonsinusoidal wave.

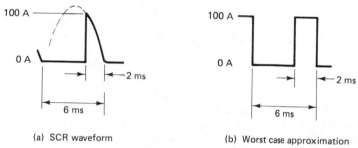

(a) SCR waveform   (b) Worst case approximation

Figure 7.2 Rectangular equivalent of SCR waveform.

After making the approximation the rms current may be determined by Eq. (7.1). Where t is pulse duration, T is pulse-repetition time, and I is peak current

$$I_{rms} = \sqrt{\frac{I^2 t}{T}} \qquad (7.1)$$

Whereas, average current of a pulse is

$$I_{AV} = \frac{It}{T} \qquad (7.2)$$

Example 7-1

Find the ratio of I_{rms} to I_{AV} of the circuit waveform shown in Fig. 7.3(a).

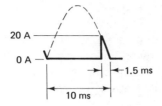

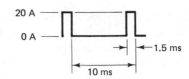

(a) Actual current waveform   (b) Approximation

Figure 7.3 Approximating the rms value of a pulse.

Solution.

1. Find the approximation [see Fig. 7.3(b)].
2. Find the I_{rms} with Eq. (7.1):

$$I_{rms} = \sqrt{\frac{I^2 t}{T}} = \sqrt{\frac{(20)^2\ 1.5}{10}} = \sqrt{60} = 7.8\ A$$

3. Find I_{AV} using Eq. (7.2):

$$I_{AV} = \frac{It}{T} = \frac{(20)\ 1.5}{10} = 3\ A$$

4. Find the ratio of I_{rms} to $I_{AV}(f_\theta)$:

$$f_\theta = \frac{I_{rms}}{I_{AV}} = 2.6$$

The ratio of the true rms current to average current is called the *form factor* (f_θ). A second method to find rms current is to multiply the average current (I_{AV}) times the form factor

$$I_{rms} = f_\theta (I_{AV}) \tag{7.3}$$

In order to find the form factor the conduction angle θ must be measured as shown in Fig. 7.4.

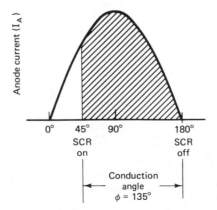

Figure 7.4 Conduction angle of an anode current.

TABLE 7.1 CONDUCTION ANGLE VS. FORM FACTOR

Conduction angle (θ)	$180°$	$160°$	$140°$	$120°$	$100°$	$80°$	$60°$	$40°$	$20°$
Form factor (f_θ)	1.3	1.4	1.6	1.8	2.0	2.3	2.7	3.5	5.0

Example 7-2

Find the rms current of a phase-controlled SCR when a dc ammeter indicates 140 A while an oscilloscope measurement shows a conduction angle of 80°.

Solution.

1. Find the form factor from Table 7.1:

$$f_\theta = 2.3$$

2. Calculate I_{rms} from Eq. (7.3):

$$I_{\text{rms}} = 2.3(140) = 322 \text{ A}$$

The SCR current rating of its anode is its repetitive-rms maximum current ($I_{T(\text{rms})}$). Caution should be taken if the given current rating of a particular SCR is not specified, specified as dc, or specified as average. If any of these three cases prevail, the current rating is average. In order to find the rms rating $I_{T(\text{rms})}$, the average rating should be divided by the form factor

$$I_{T(\text{rms})} = \frac{I_{T(AV)}}{f_\theta} \tag{7.4}$$

7.2.3 Surge-Current Rating

The surge-current rating of the SCR anode (I_{TSM}) is that amount of current an SCR can take in isolated cases. These isolated cases must be isolated by enough time for the SCR junction to cool to its rated operating temperature. The surge-current rating may be five to twenty times as large as its repetitive-current rating.

The anode surge-current rating (I_{TSM}) and the $I_{T(\text{rms})}$ should never be exceeded. If either is exceeded the SCR will be damaged and have to be replaced.

7.2.4 Maximum Forward-Blocking Voltage (V_{DRM})

The maximum forward-blocking voltage (V_{DRM}) is the principal voltage rating of the SCR. SCRs have a peak-inverse voltage rating (V_{RM}) just as diodes. But, in any SCR its V_{DRM} is lower than its V_{RM}. Hence, the circuit-peak voltage should be kept under V_{DRM}. This automatically satisfies the V_{RM} rating.

V_{DRM} is selected to be less than the forward-blocking voltage shown as the solid line in Fig. 7.1. When this rating is exceeded the SCR conducts even when there is no gate trigger. This causes unscheduled firing and prohibits control less than half-power. If V_{DRM} is greatly exceeded V_{RM} may also be exceeded. This is likely to destroy the SCR.

7.2.5 Latching and Holding Currents

The SCR acts like a latching relay. It takes a specific amount of relay-coil current to *pick* (close its contacts). After relay contacts close, a smaller coil current is sufficient to hold the relay closed. The SCR has an initial minimum value of anode current which must be flowing in order for it to stay on (latch), after gate current is turned off. This is its *latching current* (I_L). If latching current is not reached while gate current is being applied, the SCR may turn on, but it will turn off as soon as gate current is stopped.

Once the SCR is on, and remains on after gate current is removed, it is latched on. After it is latched, anode current may be lowered below the level of latching and the SCR will remain on. If anode current is lowered further, the SCR will turn off. The lower value of current, just before it turns off, is *holding current* (I_H).

Both holding and latching current have a negative temperature coefficient. At higher temperatures I_H and I_L have lower values. They have higher values at temperatures below normal.

The values of holding and latching current are usually very close to each other. These values are very small compared to $I_{T(rms)}$ at $20°C$. There are three periods when I_H and/or I_L pose a problem:

1. when the SCR is operated at low temperatures
2. when an SCR is used on a load which is far below its $I_{T(rms)}$
3. when testing SCRs with ohmmeters or limited current sources

7.2.6 Frequency and Switching Speed

SCRs are divided into two categories which reflect both frequency and switching speed ratings. These are phase-control SCRs and inverter-service SCRs. Phase-control SCRs are slower switching and have lower frequency capabilities.

There are three important ratings associated with speed: frequency at which maximum-anode current capability diminishes to zero (f_{max}), turn-on time (t_d or t_r), and turn-off time (t_q).

Frequency is given in graphical form by manufacturers as the limit where either P_{max} or I_{max} diminish to zero. As frequency increases P_{max} and I_{max} decrease. P_{max} and I_{max} are typically given at 60 Hz. At 400 Hz the typical phase-control SCR P_{max} and I_{max} capability reach zero. Inverter SCRs P_{max} and I_{max} capability typically drop to zero at 20,000 Hz.

Turn-on time is about 2 μs for both types of SCRs.

Turn-off time is about 20 μs for inverter SCRs and about 200 μs for phase-control SCRs.

7.3 SCR RATE OF CHANGE RATINGS

7.3.1 Rate of Change of Anode Current (di/dt)

When anode current initially starts in an SCR, only a very small portion of the gate-to-cathode junction is used. After a few microseconds, anode conduction spreads out evenly across the total junction. However, if current increases too fast, the initial small portion overheats, destroying the SCR.

Manufacturers set a safe value of the *rate of change of current (di/dt)* that their devices can withstand. This is called the *(di/dt)* capability of the device. Phase-control circuits with pure resistive loads are susceptible to high values of *di/dt*. Damage to SCRs may be avoided by placing inductance (L) in series with the load. The value of inductance (L) can be found with Eq. (7.5). Where (L) is inductance in henrys, *(di/dt)* max is the rate of change of current rating of the SCR in amperes per microsecond (A/μs), V_p is the peak value of voltage in volts, and 0.8 is the safety allowance for variations in tolerances

$$L = \frac{V_p}{\left(\dfrac{di}{dt}\right)_{max}}(0.8) \qquad (7.5)$$

Example 7-3

Find the value of series inductance (L) required to limit the circuit *(di/dt)* to a safe value for an SCR in a phase-control application with a 100A/μs *(di/dt)* max rating on a 240 V_{ac} line.

Solution.

1. Find V_p:

$$V_p = \sqrt{2}\,(240\ V_{ac}) = 340\ V_p$$

2. Apply Eq. (7.5):

$$L = \frac{V_p}{\left(\dfrac{di}{dt}\right)_{max}}(0.8) = \frac{340}{100}(0.8) = 2.72\ \mu H$$

7.3.2 Rate of Change of Anode-to-Cathode Voltage (*dv/dt*)

When voltage rises rapidly across an off SCR the device experiences a tempo-rary voltage *gradient*. This voltage gradient is proportional internally and places the gate (internally) at a voltage high enough to trigger the SCR. A high *rate of change of anode-to-cathode voltage (dv/dt)* causes unscheduled firing. High-frequency inverter circuits are especially susceptible to the unscheduled firing caused by high values of *dv/dt*.

Capacitance is the opposition to a change in voltage. Hence, a small value of capacitance across the SCR lowers the rate of change of voltage across the SCR. This prevents unscheduled firing. The value of capacitance (*C*) may be roughly estimated by finding the time constant (τ), then dividing it by the resistance of the SCR's load (R_L)

$$\tau = \frac{V_{DRM}}{\left(\dfrac{dv}{dt}\right)_{max}} \qquad (7.6)$$

Then

$$C = \frac{\tau}{R_L} \qquad (7.7)$$

The capacitor across the SCR charges up when the SCR is off. When the SCR is triggered on the capacitor discharges and adds to the *di/dt* presented by the original circuit. A small resistor (R_s) in series with the capacitor (see Fig. 7.5) slows the capacitor discharge. This protects the SCR from high values of *di/dt*. But, it interferes with the primary purpose of introducing the capacitor (extending the (*dv/dt*) capability of the SCR). A small diode across the series resistor (R_s) shorts it out during times of high *dv/dt*, but turns itself off during times when *di/dt* is high (see Fig. 7.5).

A rough estimate of R_s may be found with Eq. (7.8). The value found

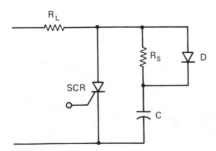

Figure 7.5 A snubber circuit to prevent unscheduled firing due to high-circuit *dv/dt*.

using this equation is usually very small. Generally, using a larger value is satisfactory. A 100-Ω resistor is a popular size.

$$R_s = \sqrt{\frac{V_{DRM}}{\left(\dfrac{di}{dt}\right)_{max}}} \tag{7.8}$$

The circuit shown in Fig. 7.5 is called a *snubber circuit*. It prevents unscheduled firing in circuits with high values of dv/dt, while preventing its own capacitance from adding to circuit di/dt.

Example 7-4

Find the values for a snubber resistor and capacitor for an inverter SCR which misfires. The SCR has V_{DRM} = 500 V, (dv/dt) max = 20 V/μs, and (di/dt) max = 40 A/μs. The SCR carries a 10-Ω load.

Solution.

1. Find the minimum time (τ) using Eq. (7.6):

$$\tau = \frac{V_{DRM}}{\left(\dfrac{dv}{dt}\right)_{max}} = \frac{500 \text{ V}}{20 \text{ V}/\mu s} = 25 \ \mu s$$

2. Find C using Eq. (7.7):

$$C = \frac{\tau}{R_L} = \frac{25 \ \mu s}{10 \ \Omega} = 2.5 \ \mu F$$

3. Find R_s using Eq. (7.8):

$$R_s = \sqrt{\frac{V_{DRM}}{\left(\dfrac{di}{dt}\right)_{max}}} = \sqrt{\frac{500 \text{ V}}{40 \text{ A}/\mu s}} = 12.5 \text{ m}\Omega$$

PROBLEMS

7.1. How can an SCR be turned on with its gate disconnected?

7.2. Explain what each of the following subscripts stand for: (a) A (b) D (c) R as a first subscript (d) R as a second subscript (e) T as a first subscript and (f) T as a second subscript.

7.3. What SCR rating is most important as all other heat-producing voltages and currents affect it.

7.4. Find the rms value of the current wave shown in Fig. 7.3 when I is 100 A, t is 2 ms, and T is 16 ms using the worst case approximation method.

7.5. Find the rms value of the current wave in Fig. 7.3 when I is 60 A, t is 1 ms, and T is 16 ms using the worst case approximation method.

7.6. Find the average value of current in the worst case approximation used to answer Problem 7.4.

7.7. Find the average value of the current in the worst case approximation used in the answer to Problem 7.5.

7.8. Find the rms current of a phase-controlled SCR when a dc ammeter indicates 100 A, while an oscilloscope shows a conduction angle (θ) is 20°.

7.9. Find the rms current of a phase-controlled SCR when a dc ammeter indicates 100 A, while a scope shows a conduction angle of 40°.

7.10. Is it ever permissible to exceed $I_{T(\text{rms})}$?

7.11. What is the nominal range of I_{TSM} for an SCR with an $I_{T(\text{rms})}$ of 20 A?

7.12. Explain the difference between surge-current and repetitive-current ratings.

7.13. What is V_{DRM}?

7.14. What is V_{RM}?

7.15. Which rating is lowest, V_{RM} or V_{DRM}?

7.16. Sketch the output-current waveform of an SCR controlling a load when V_{DRM} is slightly exceeded causing breakover with no gate trigger.

7.17. What type of a relay circuit does an SCR act like?

7.18. When is an SCR holding current greatest, at –65°C or 20°C?

7.19. List the three times when holding current may become a problem.

7.20. What is likely to happen to an SCR when (di/dt) max is exceeded?

7.21. What is likely to happen to an SCR when (dv/dt) max is exceeded?

7.22. Find the value of series inductance (L) required to limit circuit di/dt to a safe value for an SCR in a phase-controlled circuit with a 20 A/μs rating on a 480 V_{ac} line.

7.23. Sketch a snubber circuit with three components and its SCR.

7.24. Find values of R and C for a snubber when V_{DRM} = 300 V, (dv/dt) max = 100 V/μs, (di/dt) max = 100 A/μs, and R_L = 5 Ω.

7.25. List the typical f_{max}, t_d, and t_q for:
   (a) inverter SCRs
   (b) phase-control SCRs

SCR Gate Ratings and Extending SCR Ratings

8

8.1 IMPORTANT SCR GATE RATINGS

There are six important gate ratings. The following are maximum-gate ratings: *maximum-gate peak-inverse voltage* (V_{RGM}), *maximum-gate trigger voltage* (V_{GTM}), *maximum-gate trigger current* (I_{GTM}), and *maximum-gate power dissipation* (P_{GM}). When any of these are exceeded the SCR may be destroyed.

Maximum-gate peak-inverse voltage (V_{RGM}) for most SCRs is between 5 and 20 V. V_{GTM} and I_{GTM} are specified separately, and if both are used at their extreme limits at the same time, P_{GM} is certain to be exceeded. For example, *General Electric's* C35 SCR has a V_{GTM} of 10 V, and an I_{GTM} of 2.0 A. If both of these limits are placed on the C35 the power dissipation is 20 W. This is far in excess of the C35's P_{GM} of 5 W.

The two remaining important gate ratings are the smallest voltage and current required to trigger the SCR. The gate trigger must exceed both of these in order to fire the SCR. The *minimum-gate trigger voltage* is V_{GT}. The *minimum-gate trigger current* is I_{GT}. At lower temperatures both gate-trigger requirements increase. At higher ambient temperatures these requirements decrease.

8.2 EXTERNAL METHODS OF EXTENDING SCR RATINGS

The best method of avoiding the risk of exceeding SCR-maximum ratings is to use an SCR with adequate ratings. This is not always the most feasible or most economical method. Very often, SCRs with adequate ratings are not available. For example, at this printing, there is no single SCR which can carry over 5000 A or withstand over 2600 V. No single device has ratings near both of these limits. For example, there is an SCR rated at 2600 V, but it can only handle about 200 A.

SCR ratings may be extended in three ways:

1. external cooling
2. external circuitry
3. connecting SCRs in series and in parallel

8.2.1 External Cooling Methods

There are four popular methods of externally cooling an SCR:

1. mounting the SCR in cooler physical locations
2. using metal *heat sinks* on the SCR case
3. using forced air
4. watercooling

The first method of cooling the SCR may seem obvious. However, it is one of the most frequent design errors. It should be mounted in a well-ventilated location and away from other heat-producing devices—high-wattage resistors, lamps, transformers, and other high-power devices.

Another common error is to mount the SCR fuse beside the SCR. A fuse mounted under a fan will not open at rated current. Thus, it serves no end.

Figure 8.1 shows some common SCR packages and a few types of heat sinks. Heat sinks are made of metals which are good conductors of heat. Usually, they are made of copper or aluminum. They are relatively thick where they touch the device, and thinner where they contact the air. This yields as much surface area as possible to the surrounding air. The air surface may have parallel fins in order to allow convection currents of air free passage. Thermal-conductive grease is often used at the joining surfaces between the SCR and its heat sink. This promotes heat conduction from the device to its heat sink.

Figure 8.1(b), (c), and (d) shows heat sinks which have the anode of the device connected directly to the portion of the case which is connected to the heat sink. This electrically connects the anode to the heat sink. If this is undesirable, mica washers may be used between joining surfaces. Mica conducts heat well, but not electricity.

The use of fans or forced air is common in cooling high-wattage devices. Fans are directed toward the devices and their heat sinks.

Water cooling SCRs is very expensive for the initial investment. It is only used for the largest power dissipating devices.

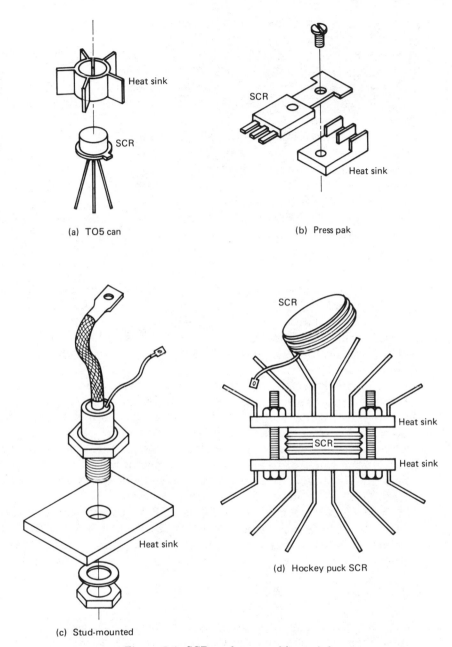

(a) TO5 can

(b) Press pak

(c) Stud-mounted

(d) Hockey puck SCR

Figure 8.1 SCR packages and heat sinks.

8.2.2 Extending SCR Ratings with External Circuitry

The RC snubber circuit shown in Fig. 7.5 is a good example of extending an SCR rating. This extends the SCR dv/dt capability.

Placing a resistor across the gate to cathode of an SCR increases V_{DRM} and dv/dt capabilities. It also decreases I_L and I_H, but requires a higher gate drive.

A negative-gate bias voltage also increases V_{DRM} and dv/dt capabilities.

A Zener diode across the gate to cathode will protect the SCR from damaging values of reverse voltage.

8.3 SERIESING AND PARALLELING SCRs

8.3.1 Connecting SCRs in Series

SCRs may be connected in series in order to extend their forward-blocking capability and reduce reverse-leakage currents. Safely connecting SCRs in series is not as simple as it first seems. Due to the differences in devices, unequal voltages would exist across two series SCRs. In order to realize the full advantages of connecting SCRs in series, a *voltage-equalizing circuit* should be used.

Voltages must be equalized during forward and reverse blocking and when the SCRs are switching off and on. Equalizing during blocking is easily accomplished using equalizing resistors in parallel with each SCR. These resistors are called *blocking-equalizing resistors* (R_{BQ1} and R_{BQ2}). They are connected as shown in Fig. 8.2. The value of these resistors is about one-tenth of the value of the blocking resistance of the SCR it is in parallel with. Then the resistance from anode to cathode of every series SCR is always within 10% of the value of any other.

During switching on and off one SCR may turn on sooner than another or switch off before the other. The OFF SCR must support the total voltage. The RC snubber circuit shown in Fig. 8.2 prevents high rate of change of voltage (dv/dt), which tends to equalize voltages during switching.

8.3.2 Triggering SCRs in Series and in Parallel

SCRs connected in series or in parallel should be triggered at the same instant. Two ways to help ensure this is to use a relatively high triggering voltage, and use a gate transformer with two identical secondaries.

The high gate voltage is called a *hard-gate drive*. This fires both SCRs faster causing less difference in turn-on time.

The *gate-trigger transformer* is used to ensure both gates are triggered simultaneously, while providing isolation (see Fig. 8.3). If the gates were

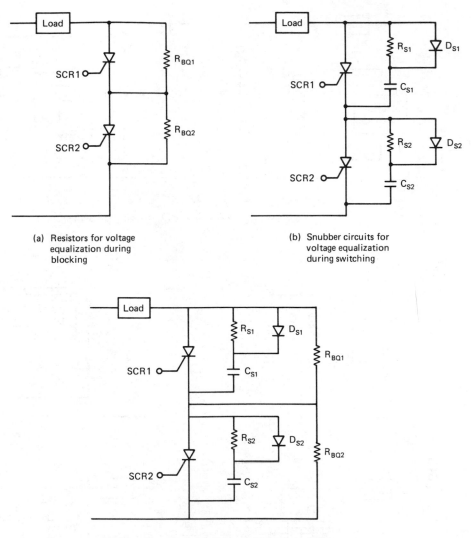

(a) Resistors for voltage equalization during blocking

(b) Snubber circuits for voltage equalization during switching

(c) Complete voltage-equalizing circuit for series SCRs

Figure 8.2 Connecting SCRs in series.

directly connected, when one fired it would load the trigger source. This decreases the trigger-source voltage and may prevent the second SCR from firing at all.

8.3.3 Paralleling SCRs

SCRs are connected in parallel in order to extend their common current capability. If SCRs are not perfectly matched, one turns on before the other, or turns off before the other. In either case the ON SCR must carry

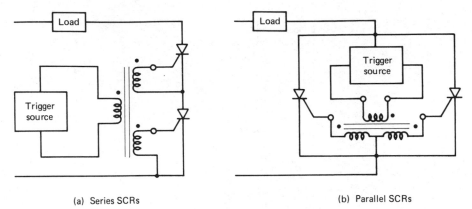

(a) Series SCRs

(b) Parallel SCRs

Figure 8.3 Simultaneous triggering of SCRs.

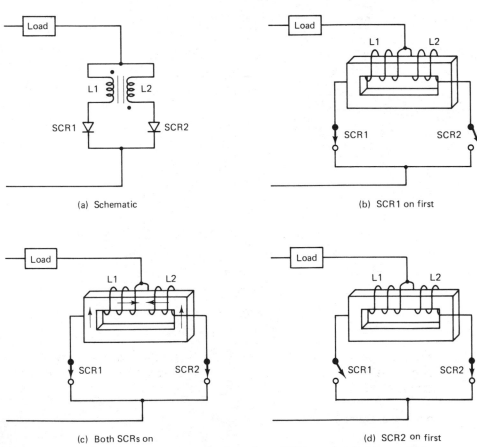

(a) Schematic

(b) SCR1 on first

(c) Both SCRs on

(d) SCR2 on first

Figure 8.4 Forced-current sharing of two SCRs connected in parallel.

the full-load current until both are switched. Matched-pair SCRs are very expensive. It is more economical to use reactors as shown in Fig. 8.4.

The reactors are wound on the same iron core. They are connected so that when current flows through both SCRs their respective fluxes cancel. Figure 8.4(b) shows SCR_1 on and SCR_2 off. The current through SCR_1 and L_1 induces a voltage across L_2 which tends to turn SCR_2 on and increase forward current through it. The absence of current through SCR_2 and L_2 increases the reactance of the L_1 coil, limiting current through SCR_1 to a safe value.

When SCRs are used in series or parallel they must be *derated* by at least 15%. For example, two 100-A SCRs in parallel may carry only 170 A (85 + 85 A).

PROBLEMS

8.1. List six important SCR gate ratings.

8.2. What is the normal range of V_{RGM} in most SCRs?

8.3. What is the best method of avoiding the risk of exceeding SCR-maximum ratings?

8.4. List three ways SCR ratings may be extended.

8.5. List four methods of externally cooling an SCR.

8.6. What material conducts heat well, but not electricity?

8.7. What device must never be cooled by a fan?

8.8. What does adding a gate-to-cathode resistor do to:
(a) *dv/dt* capability?   (b) I_H?

8.9. Sketch two SCRs in series and external circuitry to equalize voltages.

8.10. Sketch a reactive-forced current-sharing circuit using two SCRs in parallel.

8.11. What current can two 500-A SCRs connected in parallel safely carry?

SCR Phase-Control Circuits

$$9$$

9.1 INTRODUCTION

The *SCR phase control method* of controlling variable electrical power is traditionally the most popular. Switching on every cycle, it yields more intermediate settings than other methods. Its electronic circuitry is less complex. It is as smooth as a rheostat and much more efficient.

9.2 AMPLITUDE TRIGGERING

9.2.1 Amplitude Triggering of an SCR in a Half-Wave Circuit

Figure 9.1(a) shows the most basic SCR phase-control circuit. The load current path is shown in dark lines. This path consists of the source, the load, and an SCR. The maximum power from this circuit is only half of that obtained from shorting the SCR. This limits its application.

The gate-trigger circuit draws a very small current because of the large values of R_1 and R_2. Thus, very little power is consumed by the gate-trigger circuit. D_1 supplies the voltage-divider circuit (consisting of R_1 and R_2) with a half-wave dc pulse. R_1 and R_2 act to reduce this voltage and limit current to a safe value for the gate.

The wiper setting of R_2 determines the amplitude of the gate voltage (V_G). Changing V_G changes the time it takes to get to the gate-trigger voltage (V_{GT}). This controls the SCR firing and power delivered to the load.

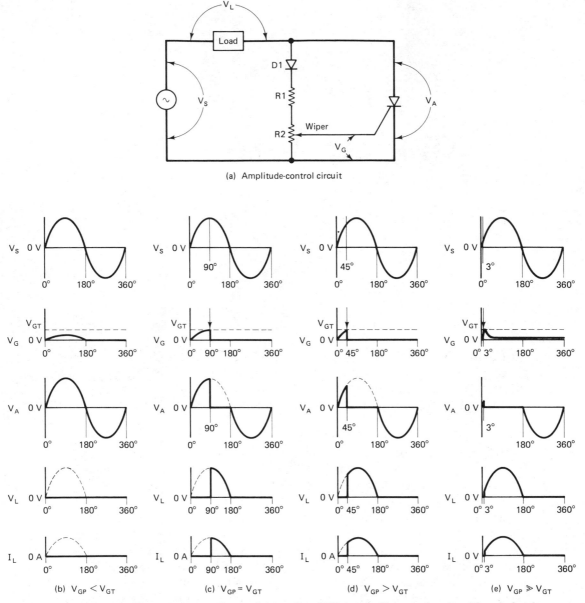

Figure 9.1 Amplitude firing of an SCR in a half-wave circuit with a dc load.

Figure 9.1(b), (c), (d), and (e) shows the circuit waveforms at various wiper settings through the circuit range of control. The first setting shows V_G peak (V_{GP}) less than V_{GT} [see Fig. 9.1(b) ($V_{GP} < V_{GT}$)]. Here, the peak voltage across the gate is never high enough to trigger the SCR. Thus, the complete source voltage (V_S) is across the SCR and no current flows through the load.

Figure 9.1(b), (c), (d), and (e) shows the circuit waveforms at various half-wave reaches its peak value at 90° and delivers half of the half-sine wave to the load.

The "shortcoming" of this control is that there is no control setting between 90° and 180°. Decreasing V_{GP} below V_{GT} turns the SCR fully off. Increasing V_{GP} above V_{GT} turns on the SCR at various times before 90°. This prohibits load control between fully off and half power. This control has limited application, but has been used to control the speed of small dc motors. It is common for many small dc motors to stop or not start with less than half power. Thus, settings below half power are unnecessary, even undesirable.

Figure 9.1(d) (V_G) shows the gate-peak voltage (V_{GP}) set at $\sqrt{2}\ V_{GT}$. V_{GP} gets to V_{GT} at 45° where it triggers the SCR. At this setting the SCR delivers about 85% of full power to the load.

When the gate triggers, V_G drops until the SCR *commutates* (turns off). This is caused by two things:

1. when the gate conducts it provides a low-resistance shunt around the lower portion of R_2 [see Fig. 9.1(a)]
2. the SCR turns on causing V_A to drop to less than 2 V

The trigger supply voltage (across D_1, R_1, and R_2) is V_A. This is desirable as prolonged heavy-gate current is unnecessary and could burn out the gate destroying the SCR. For this reason, never connect the anode of D_1 to the source end of the load. The circuit may still operate, but this presents the gate with excessive power at high settings of V_G.

Figure 9.1(e) shows the gate-voltage peak (V_{GP}) set much higher than V_{GT}. Here, V_G gets to V_{GT} at about 3° and triggers the SCR. This turns on the SCR for nearly the full half wave delivering about 99% of the full power.

9.2.2 Amplitude Triggering in a Full-Wave SCR Circuit

Figure 9.2 shows an amplitude-control circuit in a full-wave configuration. D_1 in Fig. 9.1(a) is omitted in Fig. 9.2(a) as the full-wave bridge provides a dc source. The trigger circuit consists of R_1 and R_2. Actually, R_2 alone is sufficient, except that if the R_2 wiper were set at the top (with no R_1), the full-peak line voltage would be across the gate. This would burn out the gate

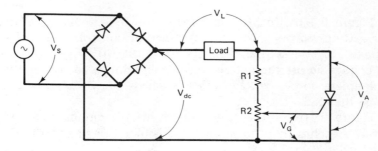

(a) Amplitude triggering of an SCR in a full-wave bridge circuit

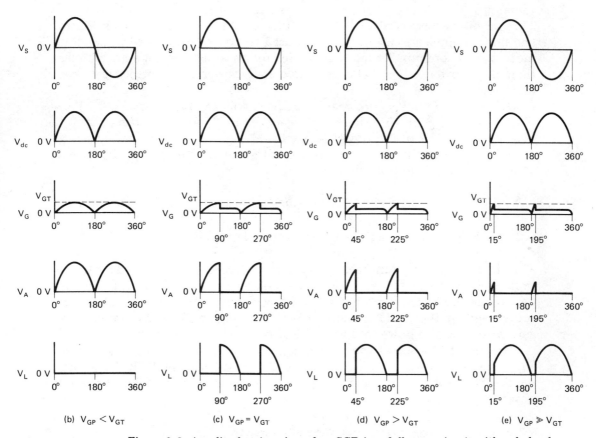

(b) $V_{GP} < V_{GT}$ (c) $V_{GP} = V_{GT}$ (d) $V_{GP} > V_{GT}$ (e) $V_{GP} \gg V_{GT}$

Figure 9.2 Amplitude triggering of an SCR in a full-wave circuit with a dc load.

and the SCR. Thus, R_1 limits the voltage across and the current through the SCR gate.

The waveforms in Fig. 9.2 are like those in Fig. 9.1 except that Fig. 9.2 shows full-wave waveforms. It is capable of delivering nearly full-line power to its load. The circuit can be used to control the speed of small dc motors. Its range of proportional control is limited to settings between half power and full power. It is capable of turning the load off, but there are no intermediate steps between off and half power.

9.3 UJT OSCILLATOR TRIGGERING

9.3.1 UJT Oscillator Triggering a dc Load

Figure 9.3 shows the UJT oscillator (see Sec. 5.3) triggering an SCR in a full-wave bridge configuration. The oscillator is the same as shown in Fig. 5.7 except for the modifications necessary for using an unfiltered source. The unfiltered source is necessary in order to commutate the SCR. It also serves to turn off the oscillator, so it starts its charge cycle fresh on each cycle of pulsating dc (V_{dc}). This synchronizes the trigger voltage with the load-SCR voltage. When the UJT fires, it sends a gate-voltage spike which triggers the SCR. After the SCR fires, the voltage across it (V_A) drops to less than 2 V. This in turn drops the voltage across R_1 and the UJT circuit [see Fig. 9.3(c), (d), or (e) (V_Z)]. The firing of the UJT also discharges C_1. The firing of the SCR and subsequent drop in V_A keeps C_1 discharged until V_{dc} increases. This allows C_1 to charge to the UJT firing point during the same time elapse on each cycle. Once R_2 is set, the UJT oscillator fires the SCR at the same time after V_{dc} goes to zero on each cycle. This delivers a constant power to the load. Changing R_2 changes the amount of power delivered to the load.

The advantages of the UJT oscillator over amplitude triggering are:

1. wider control range
2. constant voltage, thus power and current to the SCR-gate minimizing gate stress yet ensuring firing

The UJT oscillator output is ideal for triggering the SCR. The spike gives sufficient energy to trigger the SCR. Once on, the SCR is latched on. Any additional gate energy would be wasted and cause thermal burden on the SCR.

The oscillator shown in Fig. 9.3 has three modifications compared to the one shown in Fig. 5.7:

1. the addition of R_1
2. the addition of Z_1
3. the pulsating source (V_{dc}) in place of the constant dc source

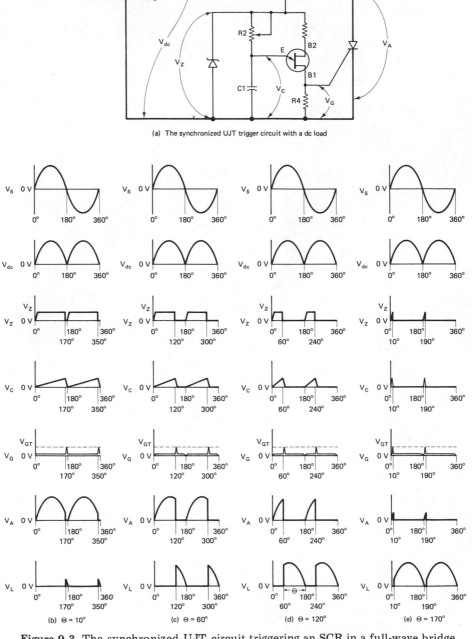

(a) The synchronized UJT trigger circuit with a dc load

(b) Θ = 10° (c) Θ = 60° (d) Θ = 120° (e) Θ = 170°

Figure 9.3 The synchronized UJT circuit triggering an SCR in a full-wave bridge circuit with a dc load.

The resistor R_1 lowers V_{dc} to a safe value for the UJT along with the other components and enables the Zener to function. The Zener (Z_1) keeps the oscillator-supply voltage at a constant value, except around the V_{dc} zero. The Zener enables the oscillator for a longer period, extending its control of the SCR closer to the V_{dc}-zero points.

Figure 9.3(b) assumes R_2 is set at a high resistance. This yields a long $R_2(C_1)$ time constant. C_1 takes about $170°$ to charge to the firing point of the UJT. The UJT fires and C_1 quickly discharges through the UJT emitter-base 1 junction and R_4. The discharge current through R_4 causes a spike of voltage (V_G) across R_4 and the gate of the SCR. This fires the SCR and turns it on. The voltage across the SCR drops and current through it increases to a value limited only by the load and V_{dc}. This current through the load causes an IR (voltage) drop across it [see Fig. 9.3(b) (V_L)].

Figure 9.3(b) shows the SCR nearly off at all times. It may be turned completely off by setting R_2 at a value high enough so that C_1 cannot charge to the UJT firing point before V_Z drops to zero.

Figure 9.3(c) assumes R_2 is set at a lower value of resistance. Here, C_1 charges to the UJT firing point at about $120°$ after $V_Z = 0$ ($\theta = 60°$). This fires the SCR for $60°$ during each cycle and delivers about $\frac{1}{4}$ power to the load.

Figure 9.3(d) shows R_2 at still a lower resistance setting. This charges C_1 to the UJT firing point at about $60°$ after $V_Z = 0$ ($\theta = 120°$). This turns the load on for about $120°$ delivering about $\frac{3}{4}$ power to the load.

Figure 9.3(e) assumes R_2 is set at zero resistance. This lets C_1 charge as V_Z rises, firing the UJT very early during every cycle of V_{dc}. This fires the SCR as soon as possible, delivering the circuit maximum power to the load (about 95%).

The circuit can turn the power completely off but cannot turn it completely on. Smooth variable control settings range from about 5% to 95% of available power. This excludes settings above zero to 5%, and above 95%.

9.3.2 UJT Oscillator Triggering an SCR with an ac Load

Figure 9.4(a) shows the same configuration as Fig. 9.3(a) except that the load is moved to the ac side of the bridge rectifier. The SCR is shown being fired at $170°$, $90°$, and $10°$ [see Fig. 9.4(b), (c), and (d)]. This supplies the load with the chopped-sine wave shown in the graphs of V_L in Fig. 9.4.

9.3.3 Firing Two SCRs with One Trigger Circuit

Figure 9.5(a) shows the synchronized UJT trigger circuit being used to alternately fire two SCRs. The pulse transformer T_2 is necessary in order to isolate the two SCRs. Transformer T_1 supplies an ac source to the full-wave centertapped-transformer rectifier with V_{S1}, V_{S2}, and a synchronized source (V_{ST}) for the trigger circuit.

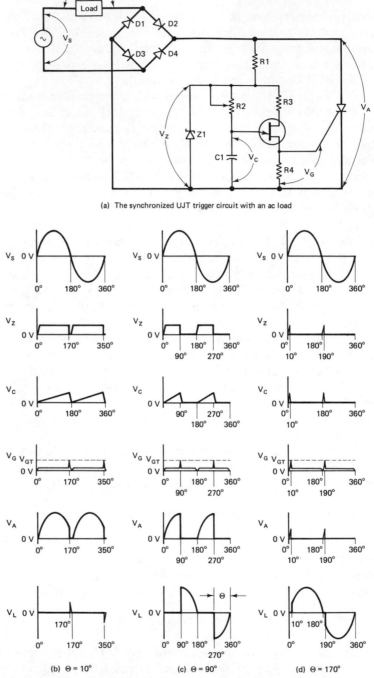

(a) The synchronized UJT trigger circuit with an ac load

(b) $\Theta = 10°$

(c) $\Theta = 90°$

(d) $\Theta = 170°$

Figure 9.4 The synchronized UJT circuit triggering an SCR in a full-wave bridge circuit with an ac load.

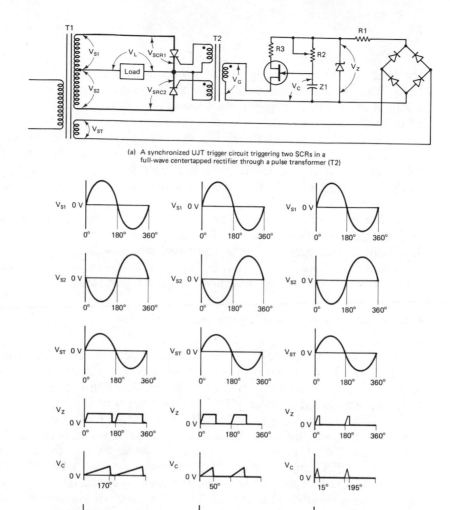

(a) A synchronized UJT trigger circuit triggering two SCRs in a
full-wave centertapped rectifier through a pulse transformer (T2)

(b) Θ = 10°   (c) Θ = 90°   (d) Θ = 165°

Figure 9.5 A synchronized UJT circuit triggering two SCRs in a full-wave centertapped-transform rectifier with a dc load.

This circuit has the advantage of less forward-voltage drop in series with the load. It has only one SCR and no diode(s) in series with the load at one time. The bridge circuit in Fig. 9.4(a) and Fig. 9.3(a) is in series with the load. This results in a higher loss when the SCR is on, and lower overall efficiency.

In addition, the SCRs in Fig. 9.5 fire alternately. Thus, each carries only half of the average current delivered to the load. This enables the use of smaller current-rated SCRs than in the single SCR used with the full-wave bridge [see Fig. 9.3(a)]. The two SCRs may be less expensive than the larger single SCR.

The main disadvantage of this circuit is the bulkiness and expense of transformer T_1. Transformer T_2 and the low-current bridge rectifier are small added expenses. Another slight disadvantage is that the trigger voltage is supplied to the reverse-biased SCR as well as the forward-biased SCR. This adds to reverse-leakage current, hence gate temperatures.

9.3.4 Possible Single-Phase SCR-Switching Configurations

Figure 9.6 shows the possible single-phase configurations where the SCR controls the load. These drawings show only the path of the principal current. The trigger circuits are omitted for simplicity. Figure 9.6(a) is the simplest configuration. It only provides half-wave to zero power.

Figure 9.6(b) may provide full ac power or half-wave dc power. With the switch open, the SCR controls dc power from zero to half-wave as in Fig. 9.6(a). With the switch closed, the negative half of the ac source is on. Then, the SCR may vary the positive half from zero to fully on. Hence, using the SCR and the diode, ac power may be varied from half to full at all intermediate points. This circuit may be used to control the speed of a *universal motor* or any resistive load.

The circuit shown in Fig. 9.6(c) has many ac applications. Both SCRs have nearly full 180° control. One SCR controls the positive half while the other SCR controls the negative half of the sine wave. This configuration is commonly known as the *SCR-ac switch*. A trigger circuit for this configuration and its waveforms are shown in Figs. 9.7 and 9.8.

The circuits shown in Fig. 9.6(d) and (e) are shown more completely in Figs. 9.3 and 9.4.

Figure 9.6(f) and (g) uses the full-wave bridge rectifier configuration. Two of the rectifiers are diodes and two are SCRs. Note that the SCRs are placed in alternate paths. That is, one is always in the path of the load current whether the ac source is positive or negative. Thus, full control of either ac or dc load may be affected. These two circuits use two SCRs, whereas Fig. 9.6(d) and (e) uses only one SCR. The two SCRs share the load current, thus, may have smaller ratings. The two SCR bridges have

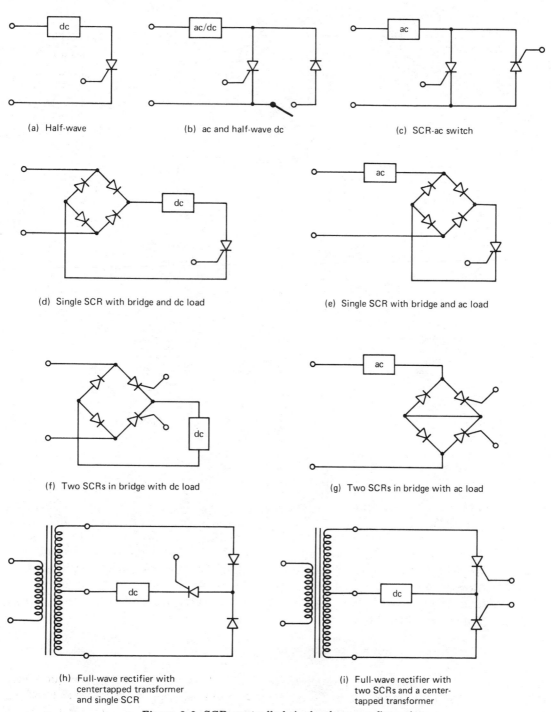

(a) Half-wave

(b) ac and half-wave dc

(c) SCR-ac switch

(d) Single SCR with bridge and dc load

(e) Single SCR with bridge and ac load

(f) Two SCRs in bridge with dc load

(g) Two SCRs in bridge with ac load

(h) Full-wave rectifier with
centertapped transformer
and single SCR

(i) Full-wave rectifier with
two SCRs and a center-
tapped transformer

Figure 9.6 SCR-controlled single-phase configurations.

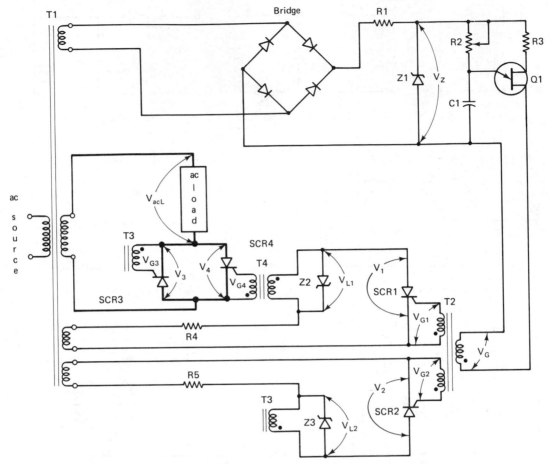

Figure 9.7 UJT triggering of the SCR-ac switch.

one less forward-diode voltage drop in series with the load. Thus, the two-SCR bridges are more efficient than the single-SCR bridge circuits.

Figure 9.6(h) and (i) both control dc loads with the full-wave centertapped-transformer configuration. Both have full-power control. The two-SCR circuit has the same advantages over the single-SCR bridge. It is more efficient and may use lower-rated SCRs. The two-SCR centertapped-transformer circuit [see Fig. 9.6(i)] is even more efficient than the two-SCR bridge [see Fig. 9.6(g)].

9.3.5 Trigger Circuit for the SCR-ac Switch

Figure 9.7 shows a trigger circuit for the SCR-ac switch. The upper-secondary winding of transformer T_1 supplies a lower voltage to the bridge rectifier and synchronized UJT oscillator. The oscillator supplies trigger spikes to

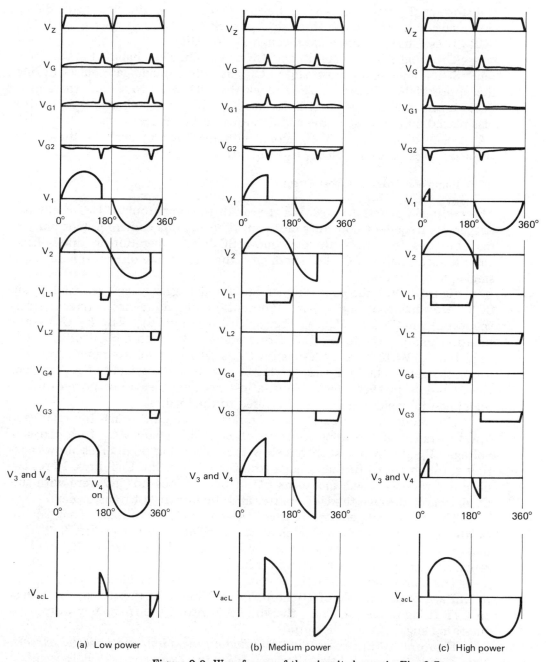

(a) Low power (b) Medium power (c) High power

Figure 9.8 Waveforms of the circuit shown in Fig. 9.7.

transformer T_2 (see Fig. 9.8, V_G) every 180°. During the first 180°, SCR$_1$ fires causing a gate pulse through T_4 to SCR$_4$. During the second 180°, SCR$_2$ fires causing a gate pulse through T_3 to SCR$_3$.

SCR$_3$ and SCR$_4$ make up an SCR-ac switch. They handle the load current. SCR$_1$ and SCR$_2$ are smaller SCRs used to provide gate pulses during the appropriate half cycles. They block the gate pulse to the alternate-load SCR during the half cycle it is reverse-biased. This eliminates the positive bias placed on the SCRs in the circuit shown in Fig. 9.5(a).

Zener diodes Z_2 and Z_3, along with resistors R_4 and R_5, limit gate voltage across SCR$_4$ and SCR$_3$ to a safe value.

9.3.6 Ramp-and-Pedestal Trigger Circuits

The *ramp-and-pedestal trigger* is a version of the synchronized UJT-oscillator trigger. Its frequency control is more sensitive and one control potentiometer may be used to control three or more SCRs for operation in three-phase circuits (see Chapter 10). For simplicity, the circuits are discussed here using single-phase circuits.

Figure 9.9(a) shows a ramp-and-pedestal trigger circuit controlling an SCR. Movement of the wiper of potentiometer R_4 varies a fixed charge (pedestal) on C_1 [see V_{ped} in Fig. 9.9(b), (c), and (d)]. The $R_2(C_1)$ combination charges C_1 to the UJT firing point faster with a high-voltage pedestal on C_1. With lower pedestals on C_1 it takes the voltage ramp across C_1 longer to get to the UJT firing point. A low pedestal setting of R_4 triggers the SCR late in every cycle. Thus, a low pedestal yields low power out. A high pedestal yields high power delivered to the load.

The diode D_1 allows C_1 to be quickly charged through the relatively low resistance of the upper portion of R_4. This charges C_1 to a pedestal voltage (V_{ped}) which is slightly below the UJT firing point. This allows current through R_2 to further charge (the ramp) until the UJT fires. Here, D_1 becomes reverse-biased and turns off. Thus, it does not interfere with the C_1 discharge through the UJT emitter and the primary of transformer T_1.

The ramp-and-pedestal is a voltage-controlled oscillator. That is, the of the control potentiometer than the oscillator shown in Fig. 9.4. This makes it suitable for use in an automatic control loop where R_4 may be replaced by a transistor or *op-amp* output.

The ramp-and-pedestal is a voltage-controlled oscillator. That is, the dc voltage on the anode of D_1 controls the time that the charge on C_1 gets to UJT firing potential. Thus, the voltage controls the frequency output or the firing angle of the SCR output.

Figure 9.10(a) shows the *modified-cosine ramp-and-pedestal circuit*. This circuit has one subtle difference from the ramp-and-pedestal circuit in Fig. 9.9(a). The capacitor charge path is through R_4 and R_2. The capacitor charge voltage is the half-sine wave, not V_Z. This causes a phase shift of the ramp voltage across the capacitor (see V_C in Fig. 10.10). A 90° phase shift

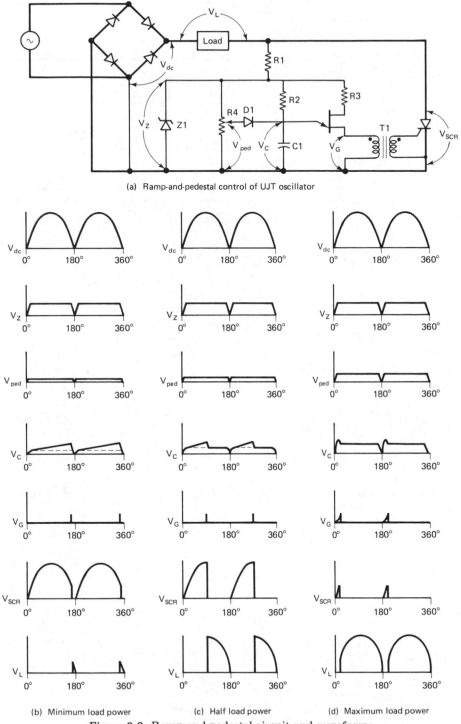

(a) Ramp-and-pedestal control of UJT oscillator

(b) Minimum load power   (c) Half load power   (d) Maximum load power

Figure 9.9 Ramp-and-pedestal circuit and waveforms.

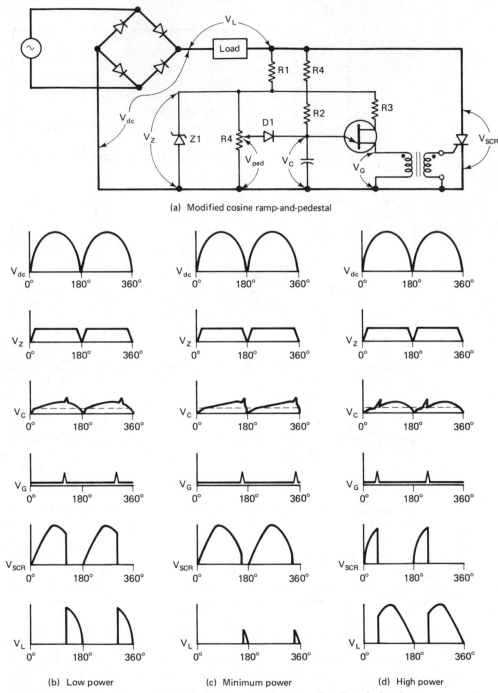

(a) Modified cosine ramp-and-pedestal

(b) Low power   (c) Minimum power   (d) High power

Figure 9.10 Modified-cosine ramp-and-pedestal circuit.

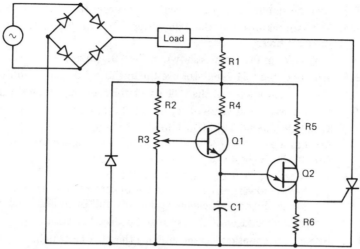

Figure 9.11 UJT trigger with a constant-current source.

of the sine wave would result in a cosine wave. The phase shift of V_C from the source is less than 90°. Yet, the circuit is called the *modified cosine*.

A disadvantage of the straight ramp-and-pedestal circuit is that there is a nonlinear relationship between V_{ped} and power delivered to the load (see Fig. 9.9). The modified cosine ramp-and-pedestal distorts V_C in the proper relationship, which makes V_{ped} directly proportional to power delivered to the load. This allows R_4 to be replaced by the output of a linear amplifier. The amplifier may be tied to any electrical output transducer completing an *automatic-control loop*.

9.3.7 UJT Trigger with a Constant-Current Source

A second circuit suitable for automatic control is shown in Fig. 9.11. Transistor Q_1 provides a constant charging current to C_1. This yields a linear ramp across C_1 and a linear relationship between the R_3 setting and UJT firing.

The R_2-R_3 voltage-divider circuit is often replaced by the output of a linear amplifier. The amplifier's input comes from a sensor circuit completing an automatic-control loop. The sensor could be a tachometer for a motor-speed control load, a thermister for a heater load, or a photoresistor for a lamp load.

PROBLEMS

9.1. What method of SCR control is traditionally the most popular?

9.2. What keeps the load current from going through R_1 and R_2 in the amplitude-control circuit in Fig. 9.1?

9.3. Can the amplitude-control circuit ever trigger the SCR from $90°$ to $180°$? Explain.

9.4. List two reasons why the gate voltage (V_G) in Fig. 9.1(b), (c), and (d) drops after the SCR fires.

9.5. Why is R_1 in the circuits shown in Figs. 9.1(a) and 9.2(a)?

9.6. Why does the UJT oscillator use the unfiltered dc shown in Fig. 9.3?

9.7. List two advantages of the UJT oscillator over amplitude triggering.

9.8. What does a lower value of R_2 in Fig. 9.3(a) do to the power delivered to the load?

9.9. Which dc-load SCR configuration in Fig. 9.6:
   (a) has the lowest efficiency due to forward-voltage drops of diodes and/or SCRs?
   (b) has the highest efficiency?

9.10. Which ac-load SCR configuration in Fig. 9.6:
   (a) has the lowest efficiency due to forward-voltage drops of diodes and SCRs?
   (b) has the highest efficiency and nearly full $0°$ to $360°$ control?

9.11. List two advantages of the ramp-and-pedestal control.

9.12. Sketch a schematic diagram showing the SCR-ac switch.

9.13. Describe what D_1 is for in Figs. 9.9 and/or 9.10.

9.14. What is the advantage of the modified cosine ramp-and-pedestal?

9.15. Why can SCR_1 and SCR_2 be rated lower then SCR_3 and SCR_4 shown in Fig. 9.7?

Three-Phase
SCR Phase Control

10

10.1 INTRODUCTION

The use of three-phase has undisputed advantages over single-phase in high-power circuits. Three-phase distribution is more efficient, more versatile, and uses less copper at the same or lower voltages. Three-phase rectifiers have a more ripple-free, unfiltered output (see Chapter 2). Three-phase rectifiers and controlled rectifiers have the advantage of using cells in parallel without the expensive equalizing reactors.

Three-phase SCR phase-control circuits often use three or six SCRs. If they were triggered simultaneously, a single UJT oscillator circuit could be used to trigger all of them. Simultaneous triggering of three-phase voltages can never deliver more than half of the available power. The three voltages are 120° out of phase [see Fig. 10.2(a)] and triggering phase A for full power at 0° triggers phase C at 120° and phase B at 240° when it is negative.

A popular method is to use three UJT oscillators. Each UJT base-bias voltage (V_{BB}) is supplied by a separately rectified phase. Then, each oscillator is synchronized to one phase voltage. The rate of change of voltage on each capacitor, across each UJT emitter-base junction, is controlled by a master dc level. Changing this level sets an equal conduction angle on each SCR, on each phase. Then, one dc level controls the full three-phase power at any point between zero and full power. This method is the ramp-and-pedestal control (see Sec. 9.3.6).

10.2 RAMP AND PEDESTAL

10.2.1 An SCR-Controlled, Half-Wave Three-Phase Rectifier

Figure 10.1 shows an important application of the ramp-and-pedestal trigger circuit. Three ramp-and-pedestal triggers are used to fire three three-phase SCRs into a dc load. Transformers T_1, T_2, and T_3 supply three UJT oscillator circuits with three-phase voltages. Phase A supplies Q_1 which is synchronized to trigger SCR_1. SCR_1 controls the amount of phase A power delivered to the load. In the same manner, Q_2 and SCR_2 are synchronized to phase B, and Q_3 and SCR_3 to phase C.

Diodes D_1, D_2, and D_3 comprise a half-wave three-phase rectifier (see Sec. 3.2). Capacitor C_1 filters this dc source providing a smooth dc across the master pedestal potentiometer (R_1). The setting of the wiper of R_1 establishes the dc pedestal across C_2, C_3, and C_4.

Resistors R_2, R_5, and R_8 are adjusted to compensate for any differences in the three UJT oscillators. When they are properly adjusted, all three oscillators fire at the same time after their respective V_Z voltages pass zero, going positive. Proper adjustment of R_2, R_5, and R_8 results in identical pulses of V_L as shown in Fig. 10.2(a), (b), and (c). Each phase of the load voltage is symmetrical with the others. Thus, R_2, R_5, and R_8 are the symmetry-adjust resistors.

Diodes D_7, D_8, and D_9 are half-wave rectifiers supplying dc to the three UJT oscillators. These dc supply voltages are flattened out by Zener diodes Z_1, Z_2, and Z_3 and resistors R_4, R_7, and R_{10}. The oscillator supply voltages are shown in Fig. 10.2 as V_{Z1}, V_{Z2}, and V_{Z3}. These half-wave supply voltages enable each oscillator when its respective SCR is forward-biased. For example, during the time period from $30°$ to $150°$, V_A is the most positive of the three-phase voltages shown in Fig. 10.2. Thus, SCR_1 is forward-biased. V_{Z1} enables the Q_1 oscillator from $0°$ to $180°$ (see V_{Z1} in Fig. 10.2).

When V_{Z1} is zero, the pedestal voltage is still placed across C_2. This forward-biases Q_1 from emitter to base, as its base is zero. Q_1 is on, but with zero voltage at B_1, it has no output. With Q_1 on, C_2 is held at the pedestal voltage until V_{Z1} comes on. Then, the Q_1 base goes positive above its emitter, turning Q_1 off. Current through R_2 charges C_2 causing the ramp on top of the pedestal (see Fig. 10.2, V_{C2}). Diode D_4 keeps the V_{C2} ramp voltage from feeding back to the Q_2 and Q_3 oscillators and also keeps C_2 from discharging through the bottom part of R_1.

When the ramp-and-pedestal voltage gets to the UJT firing point, Q_1 fires (see Fig. 10.2, V_{T4}). This causes C_2 to discharge through T_4 causing a positive spike. The spike is fed through T_4 to the gate of SCR_1, firing it. This occurs at $60°$ in Fig. 10.2(a). This turns on SCR_1 at $60°$ applying V_A across the load. SCR_1 remains on until SCR_2 fires. When SCR_2 fires, phase voltage V_B is applied across the load and to the cathode of SCR_1. This

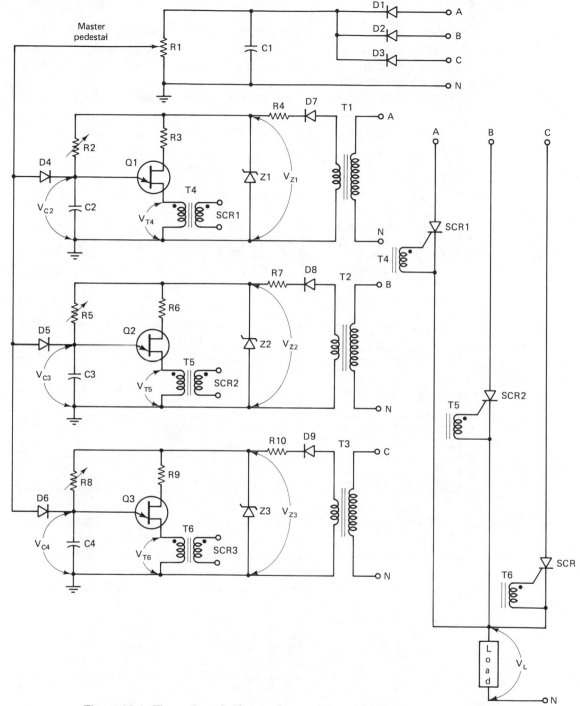

Figure 10.1 Three-phase half-wave dc control with SCRs and a ramp-and-pedestal trigger circuit.

reverse-biases SCR_1 and turns it off. Next, SCR_3 fires applying V_C to the load. This turns off SCR_2. SCR_3 remains on until SCR_1 fires. This applies V_A across the load, reverse-biasing SCR_3, turning it off. This sequence of events continues until R_1 is reset.

Figure 10.2(b) shows the pedestal at a lower value (see V_{C2}, V_{C3}, and

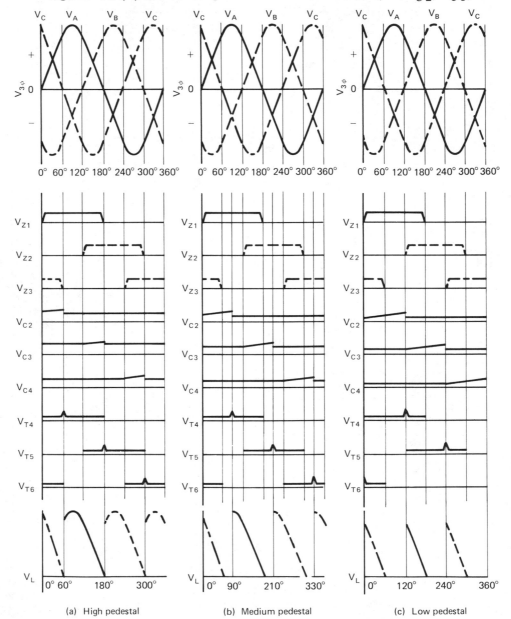

(a) High pedestal   (b) Medium pedestal   (c) Low pedestal

Figure 10.2 Waveforms of a three-phase dc supply controlled by a master pedestal and ramp.

V_{C4}). The ramp has the same charge curve or slope as in Fig. 10.2(a). It takes the ramp longer to get to the UJT firing point because it starts its charge at a lower pedestal.

The commutation of the SCRs in Fig. 10.2(b) and (c) is caused by the anode voltages going less than the neutral. The neutral voltage is graphed as the zero line in Fig. 10.2.

10.3 MASTER BIAS

10.3.1 Three-Phase, Master-Bias SCR Trigger Circuit

Figure 10.3 shows the master-bias SCR trigger circuit. Changing the master-bias control, R_1 changes the power the SCRs deliver to the load. The ramp and pedestal also uses a variable-filtered dc source to control the firing of its UJTs. It varies the pedestal under a constant slope ramp. The master-bias controls the slope of the ramp. The synchronous UJT oscillator in Fig. 10.3 varies the voltage applied across the RC charging circuit. This enables simultaneous control of more than one UJT oscillator with a single potentiometer.

Resistors R_8, R_9, and R_{10} are symmetry controls. These are adjusted to compensate for differences in UJT circuit components. They are set so that all six SCRs have equal current at any setting of R_1.

Transformers T_1, T_2, and T_3 supply the three UJTs with a phase-synchronized bias voltage. The full-wave bridge circuits furnish the UJTs with a V_Z voltage during the positive and negative half cycles. This enables each UJT to trigger its respective SCR-ac switch pair.

Figure 10.4 shows the waveforms of the master-bias circuit. The circuit is analyzed covering the firing of Q_1 and subsequent triggering of the SCR-ac switch using SCR_1 and SCR_2.

At $0°$, V_{Z1} is applied across Q_1. In Fig. 10.4(a), V_B (the master-bias voltage) is set at a relatively low level. This makes V_{C2} take about $150°$ to get to the UJT firing point. Upon Q_1 firing, the capacitor C_2 discharges through T_4 causing the spike at $150°$ in V_{T4}. This spike is fed to the gates of both SCR_1 and SCR_2. V_A forward-biases SCR_1 and reverse-biases SCR_2 at $150°$. Thus, SCR_1 turns on, and SCR_2 remains off.

At $180°$, V_{Z1} (following V_A) goes to zero. This makes the eta point (see Sec. 5.3) of the UJT base zero. The UJT turns on fully discharging C_2, but the current is so small that no spike is generated across T_4. After $180°$, V_{Z1} is applied to the UJT base. The eta point voltage rises and the UJT is off. C_2 charges until $300°$ where it fires the UJT. C_2 discharges through T_4 causing the spike at SCR_1 and SCR_2 gates. Now, SCR_2 is forward-biased and SCR_1 is reversed. SCR_2 fires causing current through load A for $30°$.

Figure 10.4(b) shows V_B at a higher level. This causes V_{C2} to charge faster. The UJT Q_1 fires sooner at $90°$ and $270°$. SCR_1 is fired at $90°$ and SCR_2 at $270°$.

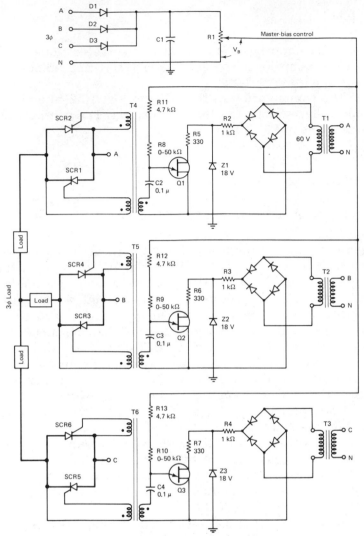

Figure 10.3 Master-bias SCR trigger circuit fires a three-phase three-wire ac load.

Figure 10.4(c) shows V_B at yet a higher level. SCR_1 and SCR_2 are fired at $30°$ and $210°$, respectively. This causes the SCR-ac switch pairs to deliver about 90% of the available power to the three-phase load.

The three-phase load may be three separate loads or any three-phase load. For example, it could be a three-phase transformer.

10.3.2 SCR Control of a Three-Phase Full-Wave Rectifier

Figure 10.5 shows UJT oscillators triggering SCRs in a three-phase full-wave bridge circuit. The trigger circuit is identical to the one shown in Fig. 10.3. The only difference is the oscillator input. The input voltages to T_1, T_2, and

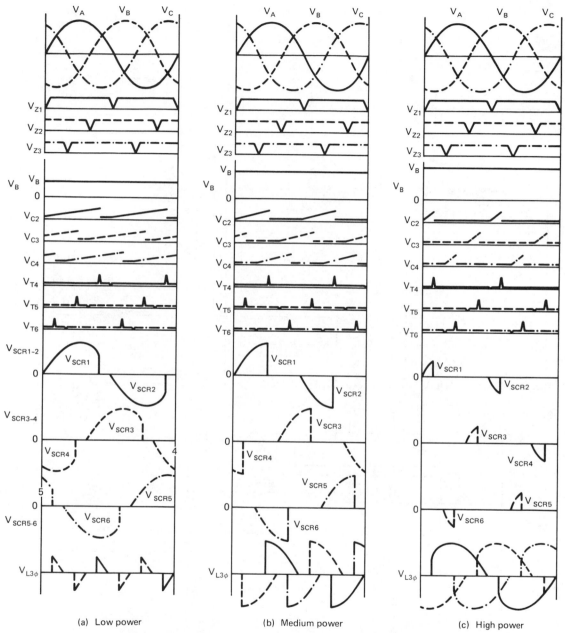

Figure 10.4 Waveforms of the master-bias SCR trigger circuit.

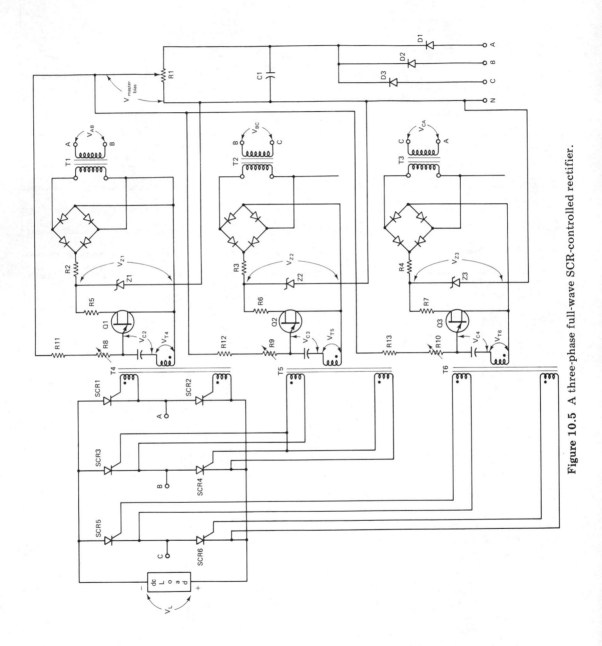

Figure 10.5 A three-phase full-wave SCR-controlled rectifier.

T_3 are V_{AB}, V_{BC}, and V_{CA}, respectively. In Fig. 10.3 the inputs are V_{AN}, V_{BN}, and V_{CN}. This synchronizes the oscillators with the three-wire source voltages used in the SCR source voltages.

Figure 10.6 shows the circuit waveforms and analysis charts. The three source voltages are shown in Fig. 10.6(a) and (b) (see $V_{3\phi}$) along with their three respective inversions. The full-wave rectifier applies the three positive peaks and the three negative peaks across the load. It inverts the negative peaks yielding six humps across the load [see V_L in Fig. 10.6(b)].

The Zener voltages (V_{Z1}, V_{Z2}, and V_{Z3}), the capacitor voltages (V_{C2}, V_{C3}, and V_{C4}), and the pulse-transformer voltages (V_{T4}, V_{T5}, and V_{T6}) are nearly the same in Fig. 10.6 as those in Fig. 10.4. The only difference is due to the setting of the master-bias voltage.

The relative-source voltages (V_{relative}) are graphed in Fig. 10.6(a) and (b). This shows the relative voltages at the terminals between the SCRs. Using these voltage graphs it can be determined when each SCR is forward-biased. For example, from $0°$ to $60°$, V_{CB} (the voltage at terminal C) is more positive than the other two source terminals A or B. Thus, the anode of SCR_6 must be more positive than its cathode (see Fig. 10.5). From $0°$ to $120°$, the voltage at terminal B (the zero line) is the most negative [see V_{relative} in Fig. 10.6(a)]. Thus, SCR_3 cathode is more negative than its anode, forward-biasing it (see Fig. 10.5).

Analysis of the most positive and negative voltages results in the line chart below V_{relative} in Fig. 10.6(a) and (b). The dark dots above each line are times when each SCR receives a gate-trigger pulse and is forward-biased. This turns each SCR on. They stay on until they are no longer forward-biased. The second line chart below V_{relative} shows when each SCR is on.

The three-phase full-wave rectifier analysis in Sec. 3.3.1 shows the current path. It is out of one source (the most positive) through its even-numbered SCR, through the load, through the odd-numbered SCR, which is tied to the most negative source terminal. Current must always pass through a pair of SCRs. One of the pair is odd-numbered and one is even-numbered. The line chart above V_L in Fig. 10.6 shows which pair of SCRs is on and when they are on.

Figure 10.6(c) shows the switch-equivalent circuits for each SCR pair. Examination of each equivalent can determine the voltage across the load when the corresponding SCR pair is on. The total of these voltages is the load voltage. For example, when SCR_3 and SCR_6 are on [see Fig. 10.6(c)], the path is from C, through SCR_6, through the load, and back through SCR_3 to B. Thus, the source, when SCR_3 and SCR_6 are on is V_{CB}. From the line chart in Fig. 10.6(a), SCR_3 and SCR_6 (3/6) are on from about $50°$ to $60°$. That portion of V_{CB} [see $V_{3\phi}$ in Fig. 10.6(a)] is the load voltage (V_L).

Figure 10.6(c) shows the switch-equivalent circuits for each SCR pair. the capacitors to UJT firing faster. Then, the SCRs fire sooner, applying

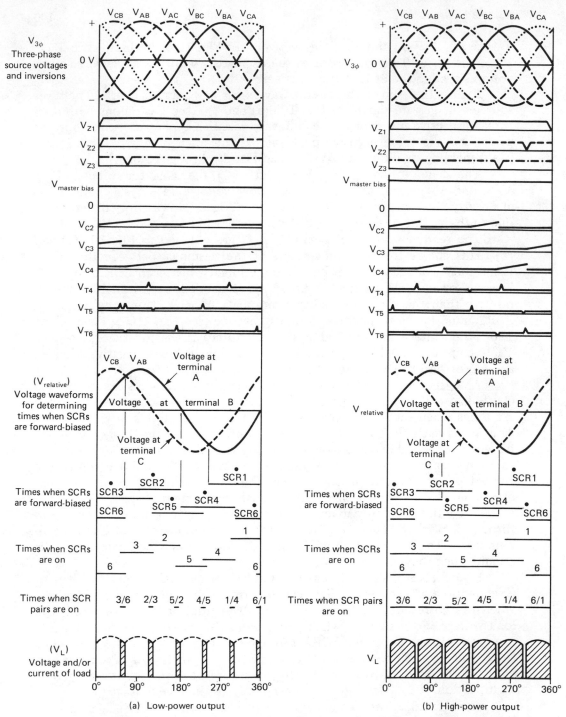

Figure 10.6 Waveform analysis of the three-phase full-wave SCR-controlled rectifier in Fig. 10.5.

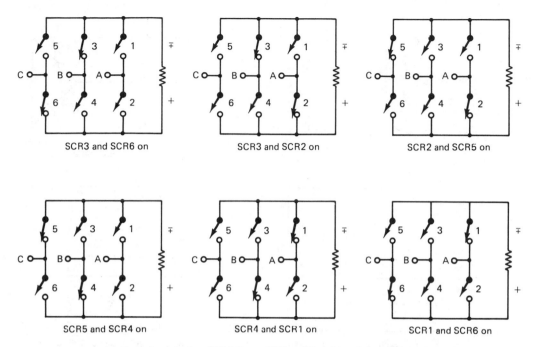

(c) Equivalent circuits of 3ϕ full-wave SCR rectifier with switches replacing SCRs

Figure 10.6 (continued)

more power to the load. Figure 10.6(c) may be used for finding V_L for Fig. 10.6(b) in the same manner as in Fig. 10.6(a).

This circuit gives a smoother, unfiltered full-power output. Each SCR need only be controlled over a range of 60° for full control. The range of full control is easier to attain than in the single-phase circuit.

10.4 THREE-PHASE CONFIGURATIONS

10.4.1 Possible SCR-Controlled Three-Phase Configurations

Figure 10.7 shows six possible three-phase SCR configurations. These circuits are simplified to show only the path of principal currents. Figure 10.7(a), (b), and (d) may be triggered by the trigger circuit shown in Fig. 10.1. Figure 10.7(e) and (f) may be triggered by the trigger circuit shown in Fig. 10.3. Figure 10.7(c) is shown with its trigger circuit in Fig. 10.5.

Table 10.1 shows the $I_{T(AV)}$ and V_{DRM} rating of the single SCR in each of the six circuits shown in Fig. 10.7.

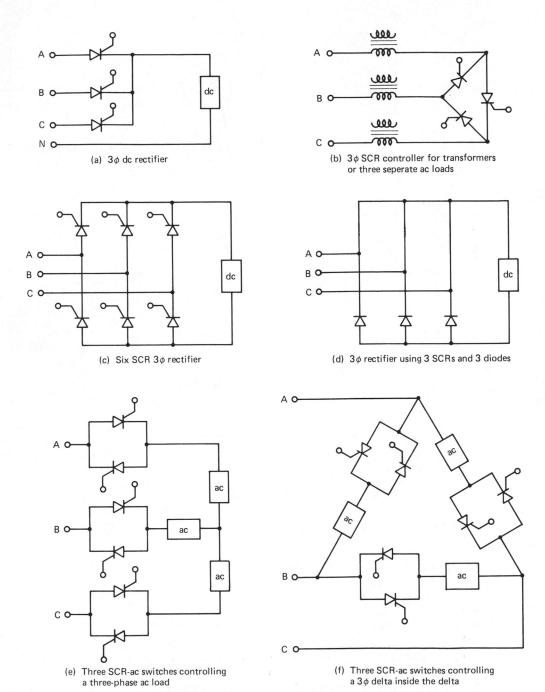

(a) 3φ dc rectifier

(b) 3φ SCR controller for transformers or three seperate ac loads

(c) Six SCR 3φ rectifier

(d) 3φ rectifier using 3 SCRs and 3 diodes

(e) Three SCR-ac switches controlling a three-phase ac load

(f) Three SCR-ac switches controlling a 3φ delta inside the delta

Figure 10.7 SCR-controlled three-phase configurations.

TABLE 10.1 V_{DRM} AND $I_{T(AV)}$ OF SCRs IN THE THREE-PHASE CIRCUITS SHOWN IN FIG. 10.7

Circuit	V_{DRM} V_s = wye source (rms)	$I_{T(AV)}$ I_L = line current
(a)	$1.732 \, V_s$	$0.276 \, V_s/R_L$
(b)	$1.732 \, V_s$	$1.732 \, I_L$
(c)	$1.732 \, V_s$	$0.552 \, I_L$
(d)	$1.732 \, V_s$	$0.552 \, I_L$
(e)	$1.732 \, V_s$	$0.276 \, I_L$
(f)	$1.732 \, V_s$	$0.276 \, I_L$

PROBLEMS

10.1. If T_4 were accidentally tied to SCR_2, and T_5 to SCR_1 in Fig. 10.1, what would be the output for the setting shown in Fig. 10.2(b) (show with a sketch of V_L)?

10.2. Explain the differences in the master-bias and the ramp-and-pedestal controls.

10.3. Explain why the rectifier shown in Fig. 10.7(d) is more efficient than the one shown in Fig. 10.7(c).

10.4. Sketch a three-SCR ac-switch controlling a three-phase delta load with the ac switches in the line, not in the delta as shown in Fig. 10.7(f).

10.5. How long (in degrees) is each SCR forward-biased in Fig. 10.5?

Electronic Inverters

11

11.1 INTRODUCTION

An inverter changes dc to ac. Inverters are used in ac-induction motor control, induction heating systems, and in many applications where only dc is available and ac is required. Many of these latter applications are in aircraft, ships, and land vehicles. Inverters are also used in dc converters and cycloconverters.

A dc converter changes dc at one voltage to dc at another voltage. This is done by first using an inverter (changing dc to ac), changing the voltage value with a transformer, and finally, rectifying the ac to dc at the desired voltage.

A cycloconverter converts ac at one frequency to ac at another frequency. The initial ac is first rectified. This dc is fed to an inverter where the desired frequency is produced.

11.2 TRANSISTORIZED INVERTERS

11.2.1 A Single-Phase Transistor Inverter

Figure 11.1 shows a simplified schematic of a transistor inverter. A square-wave oscillator uses dc to generate a square-wave source. This source must supply enough base current to saturate transistors Q_1 and Q_2.

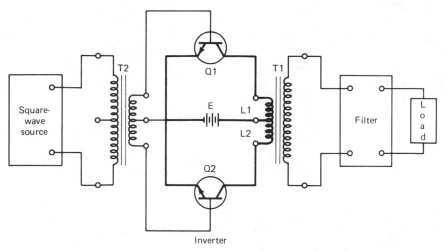

Figure 11.1 A simplified schematic of a square-wave inverter.

The square wave is fed through transformer T_2 to the bases of Q_1 and Q_2. First, Q_1 is switched into saturation. The dc source (E) causes current flow through L_1 and Q_1 back to ground.

When the square-wave source goes negative it switches Q_1 off and Q_2 on. This causes current flow through L_2, Q_2, and back to ground.

Then, Q_2 switches off and Q_1 is switched on again. As Q_1 and Q_2 are alternately switched, current through the primary of T_1 (L_1 and L_2) alternates. This induces a square-wave voltage across the secondary of T_1. This square wave is filtered, making a nearly perfect sine wave, and then it is applied to the load.

Most inverters have a square-wave output. The main reason for this is efficiency. For example, biasing Q_1 and Q_2 half on with a sine wave applied, would result in a sine-wave output. But then, the average voltage across the transistors would be equal to the voltages across L_1 and L_2. The transistors would be operating like inefficient rheostats instead of switches (see Sec. 1.1.2).

Transistors have certain advantages over SCRs in inverter circuits. They have a lower forward-voltage drop than SCRs, and they are more easily turned off.

Yet, SCRs are available with higher current and voltage capabilities and they have a higher gain. It takes more power to turn on a transistor than it does an SCR. The added drive makes transistors less efficient.

Figure 11.2 shows the complete inverter that was shown in simplified form in Fig. 11.1. This circuit shows three complete stages and the ac load.

The inverter stage shows three additional components: D_1, D_2, and C_2. Turning off current in the transformer primary causes high *transient* voltages to be generated in the off-half of the transformer primary. Diodes D_1 and D_2 offer a path for transient current to flow. This protects the

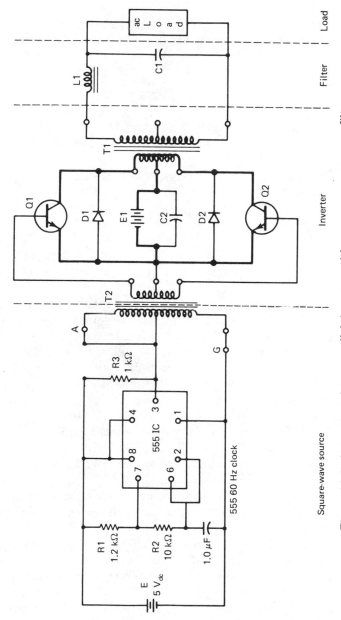

Figure 11.2 A two-transistor parallel inverter with square-wave source, filter, and load.

117

transistors from high-reverse voltages. Capacitor C_1 acts on this same inductive energy. It provides a shunt for high transients and stores some energy during switching.

The square-wave source consists of a 555 *integrated circuit* (IC) and frequency determining components. The frequency of this source is determined by

$$f = \frac{1.46}{(R_1 + 2R_2)\,C} \tag{11.1}$$

The frequency may be changed by changing C or R_1 and R_2. R_1 and R_2 must be kept at the same ratio shown or the output becomes rectangular instead of square.

The LC filter rounds off the square wave so it more nearly approximates a sine wave.

11.2.2 General Inverter Configurations

The dark-lined circuit shown in Fig. 11.1 should look familiar. If the transistors were diodes and source and load interchanged, it would become a full-wave rectifier.

Most inverter circuits may be inverted into rectifiers. Replace the dc source with a dc load, the ac load with an ac source, the controlled switches with diodes, and an inverter becomes a rectifier.

Figure 11.3(a) shows the inverted half-wave rectifier. This circuit is called a *chopper* as its output is a chopped dc. It is commonly used to control dc-vehicle motor speed. For full power, the electronic switch is on all the time. Often, a mechanical switch is connected across the electronic switch for full power. At half-speed the electronic switch is on about half the time. For lower speeds it is on for shorter periods and off for longer periods.

The operation of the circuit shown in Fig. 11.3(b) is covered in Sec. 11.2.1.

Figure 11.3(c) shows a bridge configuration. Switch pairs S_1 and S_4 must be on and off together and S_2 and S_3 must be synchronized. If S_1 were on with S_3, the source would be shorted.

With S_1 and S_4 on, current flows from $+E$ through S_1, the ac load, S_4 and back to $-E$. With S_2 and S_3 on, current flows from $+E$ through S_2, the ac load, S_3, and back to $-E$. The ac peak-to-peak output is $2E$.

Three-phase inverters are shown in Fig. 11.3(d) and (e). The half-wave inverter shown in Fig. 11.3(d) must be fired into a four-wire wye load. The load could be converted to a delta using transformers. The peak-to-peak voltage is E.

The circuit shown in Fig. 11.3(e) has a peak-to-peak voltage output equal to $2E$. By staggering the six electronic switch-on times, an output

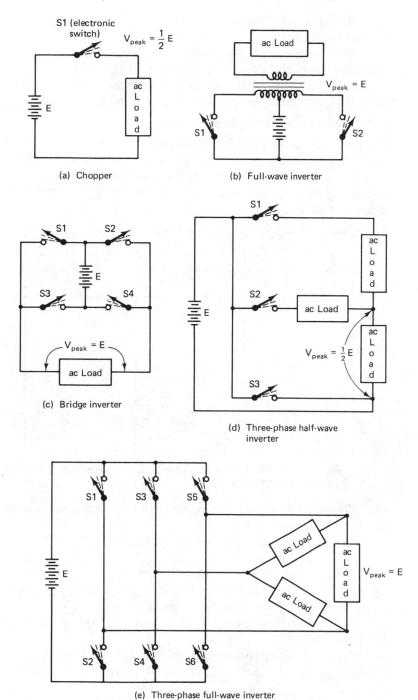

(a) Chopper

(b) Full-wave inverter

(c) Bridge inverter

(d) Three-phase half-wave inverter

(e) Three-phase full-wave inverter

Figure 11.3 General inverter configurations.

more closely resembling a sine wave may be formed. The circuit may be used with a three-wire wye or delta load.

Both circuits are covered in detail using a ring-counter trigger in the following sections. Digital circuits are often used to trigger these circuits, but they are beyond the scope of this text. A basic digital text should be consulted. A *mod-9* counter has outputs available to trigger either circuit.

11.2.3 The Two-On, Three-Phase Full-Wave Inverter

Figure 11.4 shows an FET three-phase full-wave inverter. It is a two-on as two FETs are always on. Section 11.3.2 covers the three-on. This inverter is well suited to drive a three-phase induction motor, as it is capable of a variable-frequency output. This can be used to control the speed of an induction motor.

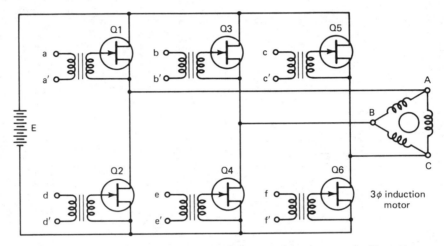

Figure 11.4 An FET three-phase full-wave inverter controlling the speed of an induction motor.

Figure 11.5 shows the firing schedule which is necessary for the sine-like outputs shown in Fig. 11.6.

Figure 11.5(a) shows the state of the FETs from $0°$ to $60°$. Q_1 connects the positive end of E ($+E$) to A, and Q_4 connects the negative end of E ($-E$) to B. Thus, from $0°$ to $60°$ the voltage from terminal A to terminal B (V_{AB}) equals $+E$. The voltages from A to C (V_{AC}) and from C to B (V_{CB}) are both equal to $+\frac{1}{2}E$. Figure 11.6 shows the inverse of V_{AC} and V_{CB}, which is V_{CA} and V_{BC}, respectively. Hence, V_{CA} and V_{BC} are equal to $-\frac{1}{2}E$ from $0°$ to $60°$ as shown.

Figure 11.5(b) shows Q_1 and Q_6 on. This makes V_{AC} equal to $+E$, hence V_{CA} is $-E$. Both V_{AB} and V_{BC} are $+\frac{1}{2}E$ from $60°$ to $120°$ (see Fig. 11.6).

By analyzing Fig. 11.5(c), (d), (e), and (f) and graphing the results, the three composite waveforms shown in Fig. 11.6 are formed.

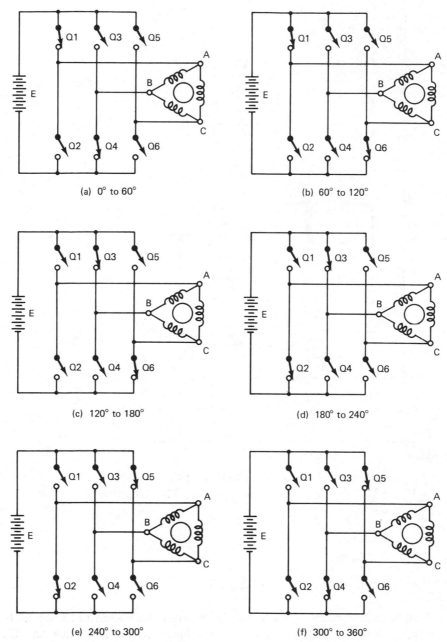

Figure 11.5 Simplified schematics of the inverter shown in Fig. 11.4, showing the transistor firing schedule for the output waveform analysis (see Fig. 11.6).

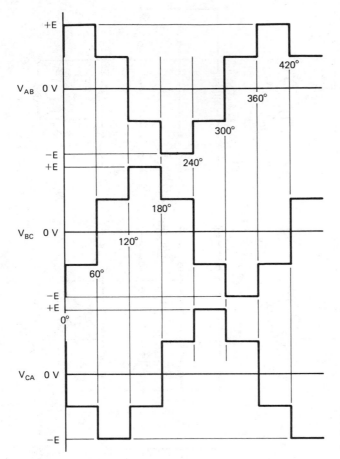

Figure 11.6 The output-voltage waveforms of the three-phase full-wave inverter shown in Fig. 11.4.

The trigger circuit for the three-phase full-wave FET inverter is shown in Fig. 11.7. It is an SCR ring counter. The trigger circuit is also an inverter in its own right.

11.3 SCR INVERTERS

11.3.1 The Ring-Counter Trigger

SCR inverters must use a special method to commutate the SCR. Once the gate trigger turns the SCR on, voltages on the gate have no effect on conduction. The three basic methods of attaining commutation in inverters are:

1. firing another electronic switch to cause a zero current in the first SCR
2. firing the SCR into a tuned-resonant circuit
3. using an external negative pulse on the SCR anode

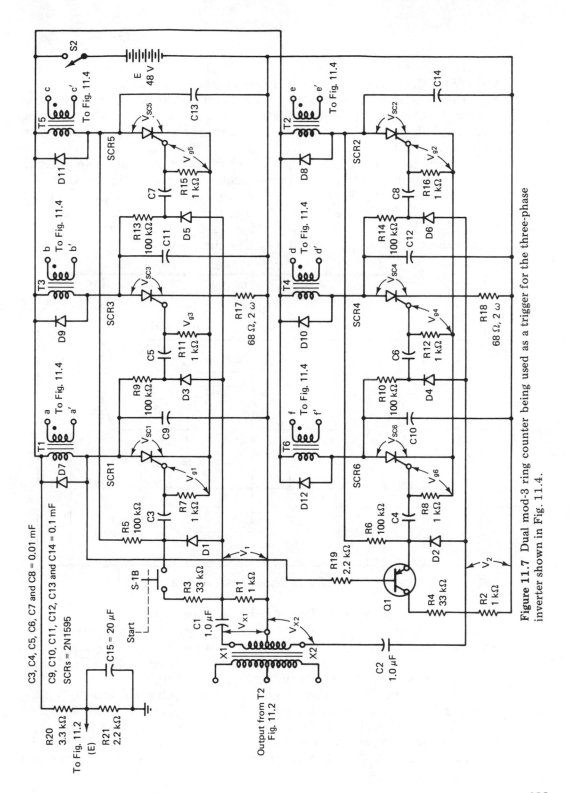

Figure 11.7 Dual mod-3 ring counter being used as a trigger for the three-phase inverter shown in Fig. 11.4.

The ring counter shown in Fig. 11.7 is of the first type. Commutation of one SCR depends on another SCR being turned on.

The frequency output of the ring counter is established by the frequency of the square-wave oscillator shown in Fig. 11.2. Varying the frequency of the oscillator varies the speed of the three-phase induction motor shown in Fig. 11.5.

The circuit in Fig. 11.7 consists of two mod-3 ring counters. In order to cover its operation, a single mod-3 ring counter is shown in Fig. 11.8.

When S2 in Fig. 11.8 is on, capacitors C_3, C_9, C_5, C_{11}, C_7, and C_{13} charge to E. C_3 charges through T_5, R_5, R_7, and R_{17}. C_9 charges through T_1. C_5 charges through T_1, R_9, R_{11}, and R_{17}. C_{11} charges through T_3. C_7 charges through T_3, R_{13}, R_{15}, and R_{17}. C_{13} charges through T_5.

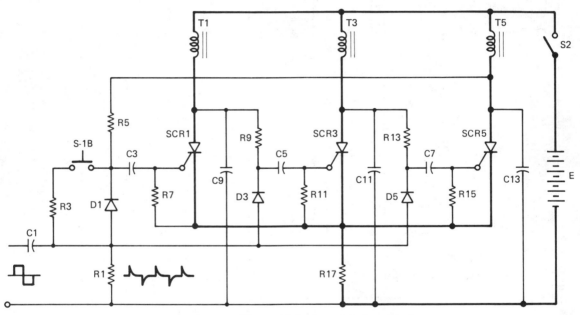

Figure 11.8 A mod-3 ring counter.

A square-wave input voltage is fed to C_1 across R_1. C_1 and R_1 *differentiate* the square wave into spikes as shown in Fig. 11.8. The spikes would normally provide triggers to all three SCRs through diode-capacitor pairs D_1–C_3, D_3–C_5, and D_5–C_7. However, the positive charge of E on C_3, C_5, and C_7 reverse-biases their respective diodes D_1, D_3, and D_5.

When the start switch (S-1B) is closed, C_3 is discharged through R_3, R_1, R_{17}, and R_7. This allows a positive spike through D_1 and C_3 to trigger SCR$_1$. Then, C_5 discharges through R_9, SCR$_1$, and R_{11}. With C_5 discharged, D_3 is no longer reverse-biased. This enables the next positive trigger to fire SCR$_3$. When SCR$_3$ fires, C_{11} discharges through SCR$_3$ and R_{17}. This yields a high positive-voltage spike across R_{17} causing SCR_1 cathode to go positive. This spike momentarily reverse-biases SCR_1 and commutates it. C_7 also discharges through R_{13}, SCR_3, and R_{15} while SCR_3 is on. This enables SCR_5 to be triggered by allowing the third positive spike through D_5. When SCR_5 fires it discharges C_{13} and C_3. Discharging C_{13} causes a high positive voltage across R_{17}, which commutates SCR_3. The discharge of C_3 enables the next positive trigger to fire SCR_1 again. When SCR_1 fires it commutates SCR_5 and enables SCR_3. The counter keeps repeating. It turns on SCR_1, then SCR_3, then SCR_5, and then SCR_1 again. The circuit waveforms are shown in Fig. 11.9(a).

The voltage waveforms shown in Fig. 11.9 are labeled in Fig. 11.7. Voltages V_{x1} and V_{x2} are from the secondary of T_2 in Fig. 11.2. V_{x1} is the inverse of V_{x2}. Both of these square waves are differentiated and fed to the anodes of the six enabling diodes. V_1 is fed to D_1, D_3, and D_5. These are enabled in sequence as discussed in the last paragraph with reference to Fig. 11.8.

The degree-time reference in Fig. 11.9 uses the output voltage of the inverter shown in Fig. 11.4. Thus, during one cycle of the output, six SCRs are triggered consecutively, one every 60°. This provides six FET gate pulses. Each pulse is 120° in length, but the pulses overlap by 60°. There are always two FET gate pulses on, but the on-pair changes every 60°.

Using the gate pulses shown in Fig. 11.9 ($V_{aa'}$, $V_{bb'}$, $V_{cc'}$, $V_{ff'}$, $V_{dd'}$, and $V_{ee'}$), the firing schedule of the Q_1 through Q_6 FETs may be found. This agrees with the firing schedule shown in Fig. 11.5.

The bottom mod-3 ring counter shown with SCR_6, SCR_4, and SCR_2, in Fig. 11.7 operates the same as the one discussed in Fig. 11.8. There are some additional components. Resistor R_{19} provides bias for Q_1. The emitter and base of Q_1 is normally at $+E$. Its base is tied to E through R_{19} and T_1. This keeps Q_1 off until SCR_1 fires. When SCR_1 fires, its anode voltage drops and forward-biases Q_1. This enables SCR_6 while SCR_1 is on. This is to disable SCR_6 until SCR_1 fires. If the lower ring counter were allowed to start at random, SCR pairs such as SCR_1 and SCR_2 may fire during the same 120° period. This would turn on Q_1 and Q_2 in Fig. 11.4 and provide a direct short across E.

When SCR_1 enables SCR_6, the FET firing schedule will be correct. This avoids the turn-on corresponding FET pairs, which causes a direct short across the dc source (E) (see Fig. 11.4).

Another possible problem is that if E is applied simultaneously with the square-wave oscillator, all SCRs may turn on before the gate-disabling capacitors (C_3 through C_8) can charge to E. This is prevented by using the 48 V E in Fig. 11.7 for the 20 V E in Fig. 11.2. The 48 V are applied across the R_{20}–R_{21} voltage divider shown in the upper left-hand corner of Fig. 11.7. Before the 20 V reach the oscillator, C_{15} must be charged. This causes a time delay which is much longer than the time required to charge the disabling

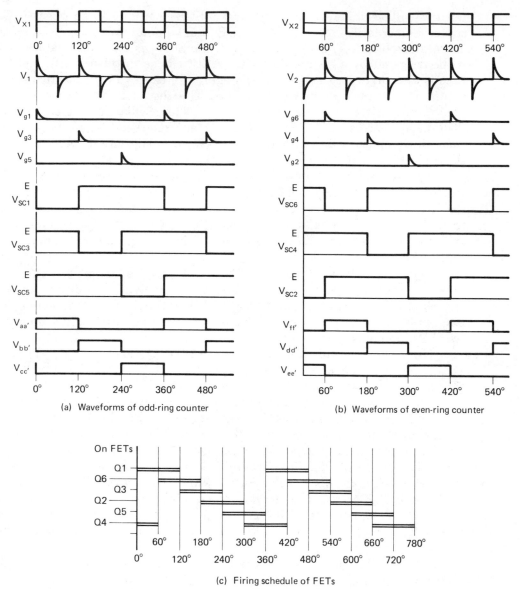

(a) Waveforms of odd-ring counter

(b) Waveforms of even-ring counter

(c) Firing schedule of FETs

Figure 11.9 Waveforms of ring-counter trigger circuit.

capacitors. This avoids simultaneous turn-on of all SCRs and the consequent short circuit across E (see Fig. 11.4).

Diodes D_7, D_8, D_9, D_{10}, D_{11}, and D_{12} are *freewheeling* diodes. They prevent their respective SCRs from being held on by the inductive load of the primaries of T_1, T_2, T_3, T_4, T_5, and T_6.

Inductance opposes a change in current. For example, when SCR_1 commutates, the inductance of T_1 primary produces current attempting to keep

SCR_1 on. Diode D_7 shorts out this inductive current. This prevents SCR_1 from freewheeling (remaining on).

The three-phase inverter described in Figs. 11.4 through 11.9 is called a *two-on, six-step* three-phase inverter. The two-on always has two primary-current cells (in this case FETs) on. The next section describes the *three-on* inverter.

11.3.2 The Three-On, Six-Step Three-Phase Inverter

Figure 11.10 shows a three-phase full-wave inverter. Essentially, it is the same as the circuit shown in Fig. 11.4. The FETs have been replaced with SCRs and the delta load has been replaced by a wye in order to demonstrate industrial usage. The primary-current cells could be bipolar transistors, FETs, SCRs, or even ignitrons. The load could be either delta or wye. The significant difference is the firing schedule. In the three-on inverter each cell is on for 180° instead of 120°. This is an advantage of the three-on as switching cells less frequently lowers the demand on individual cells. A disadvantage of the three-on is that *shoot-through* is more likely to occur. Shoot-through is when a series pair of cells is on simultaneously. Series pairs are $SCR_{1,2}$, $SCR_{3,4}$, or $SCR_{5,6}$. If any series pair is on at once the dc source (E) suffers a direct short.

Figure 11.11(a) shows the switch equivalent of the three-on inverter for each of the six switching steps during one cycle. Figure 11.11(b) shows each equivalent without the switches. This shows that two of the three windings in the wye load are always in parallel and one is in series with the parallel pair. Thus, the impedance of the parallel pair is half the impedance of one winding or $\frac{1}{3}$ of the total impedance. Applying the *voltage-divider formula*, the parallel pair drops $\frac{1}{3}E$ while $\frac{2}{3}E$ is dropped by the single-series winding. This assumes that all three wye windings are equal in impedance. The output voltages V_{AN}, V_{BN}, and V_{CN} are similar to those shown in Fig. 11.6 but peak at $\frac{2}{3}E$ instead of E.

11.3.3 SCR Flasher Circuit

The circuit shown in Fig. 11.12 is commonly used for a two-lamp flasher. More important, its operation is much like the popular *McMurray-Bedford* inverter shown in Fig. 11.13.

The trigger circuit for the inverter in Fig. 11.12 triggers SCR_1, then SCR_2 alternately. SCR_1 in Fig. 11.1(b) is triggered first. This allows C to charge to E through R_2. When SCR_2 fires, C discharges through SCR_2 and SCR_1. The inverse current through SCR_1 commutates it. Next, C charges in the reverse direction shown in Fig. 11.2(b). Then, when SCR_1 is fired, C discharges into the cathode of SCR_2 and commutates it. The process keeps repeating. The frequency output is totally dependent on the trigger input.

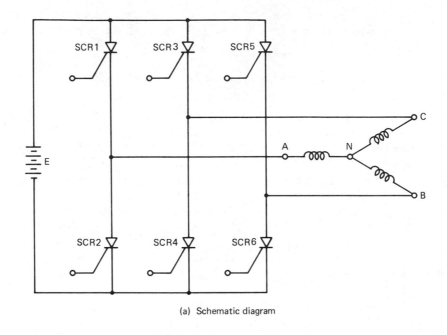

(a) Schematic diagram

Note: shaded areas indicate ON times

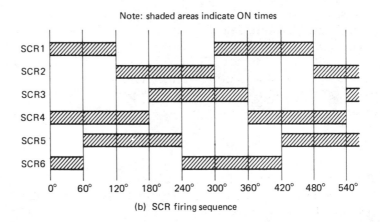

(b) SCR firing sequence

Figure 11.10 The three-on, six-step three-phase inverter and firing sequence.

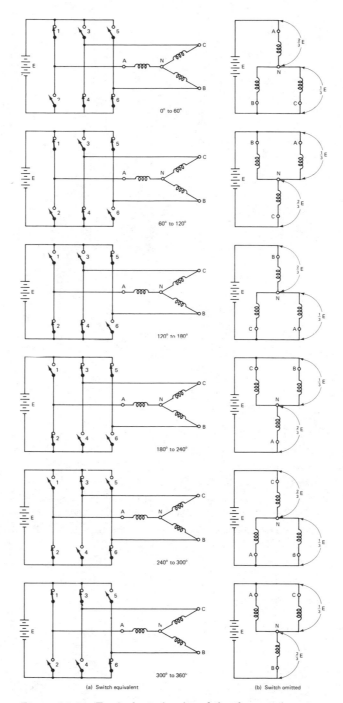

(a) Switch equivalent   (b) Switch omitted

Figure 11.11 Equivalent circuits of the three-on inverter.

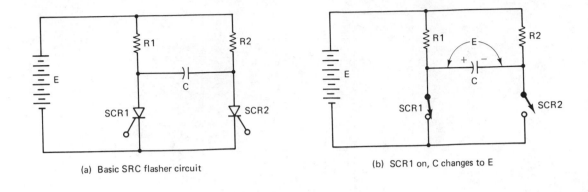

(a) Basic SRC flasher circuit

(b) SCR1 on, C changes to E

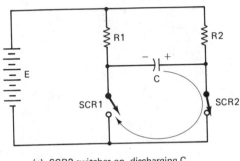

(c) SCR2 switches on, discharging C,
and commutating SCR1

Figure 11.12 The operation of the basic SCR flasher circuit.

The McMurray-Bedford inverter shown in Fig. 11.13 is commutated by C_1 in the same manner as the flasher circuit in Fig. 11.12. Its peak output is E which is fed to the ac load through T_2. Inductance L aids in commutation much like the discharge through the cathode resistor R_{17} in Fig. 11.8. Diodes D_1 and D_2 shunt the tendency of inductive currents to oppose a change in current and keep the SCRs from commutating.

The square wave is fed through T_1 and is differentiated by the twin sets (C_2, R_1 and C_3, R_2). Both the positive and negative spikes are fed to both gates. However, the positive spike to SCR$_1$ gate is the negative to SCR$_2$ gate and vice versa; Thus, the SCRs are fired alternately.

PROBLEMS

11.1. List three applications of inverters.

11.2. Sketch a block diagram of a dc converter. Use and label three blocks.

11.3. Sketch a block diagram of a cycloconverter. Use and label three blocks.

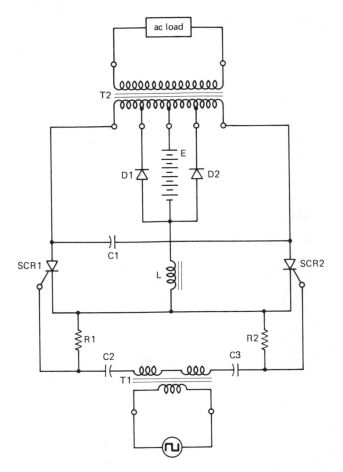

Figure 11.13 The McMurray-Bedford inverter.

11.4. Why do electronic inverters have square-wave outputs?

11.5. Explain what the freewheeling diodes in Fig. 11.2 are for.

11.6. What would happen in Fig. 11.3(e) if S_5 and S_6 were on?

11.7. List two inverter circuits which commutate SCRs using the discharge of a capacitor.

11.8. Explain why the output voltage V_{BC} in Fig. 11.6 is lower in amplitude during the period from $60°$ to $120°$ than it is between $120°$ and $180°$.

11.9. Describe the function of C_{11} in Fig. 11.8.

11.10. Describe what would happen in Fig. 11.13 if L were shorted.

11.11. Sketch the graphs of voltages V_{AN}, V_{BN}, and V_{CN} for the three-on three-phase inverter. Label the horizontal axis from $0°$ to $360°$ in $60°$ steps and the vertical axis from $+\frac{2}{3}E$, $+\frac{1}{3}E$, 0 V, $-\frac{1}{3}E$, and $-\frac{2}{3}E$ (see Fig. 11.11).

Zero-Voltage Switching

12

12.1 INTRODUCTION

Zero-voltage switching is used where the combination of high power and high frequency requires precise control. Phase control of high frequency/high power results in very strong radio-frequency interference (RFI). Phase-control outputs are close to square wave (see Sec. 7.2.2 and Fig. 7.2). These waveforms emit strong harmonics. High-frequency emissions may be harmful to people and equipment and interfere with radio and television communications. Thus, they are illegal. RFI limits are enforced by the Federal Communications Commission (FCC).

The strength of these RFI harmonics diminishes as the frequency of the harmonic increases. Even so, some 60 Hz phase controllers exceed FCC limits. Thus, base frequencies higher than 60 Hz are much more susceptible to excessive RFI. Zero-voltage switching eliminates excessive RFI.

Zero-voltage switching is used mainly with induction heating devices. The induction coil surrounds the material to be heated. The energy is fed to the coil which induces it into the material. This induction is more efficient at higher frequencies—some as high as 20 kHz.

Zero-voltage switching circuitry is covered after some system limits and control definitions are discussed.

12.2 DEFINITIONS AND SYSTEMS

12.2.1 Zero-Voltage Duty Cycle

The *zero-voltage duty-cycle time* (t_d) is a multiple of the cycle time of the controlled source power. It is shown to be ten times as long as the *source-cycle time* (t_s) in Fig. 12.1. The ratio of t_d to t_s plus one is the number of *control steps* (N_{cs}), or

$$N_{cs} = \frac{t_d}{t_s} + 1 \qquad (12.1)$$

TABLE 12.1 PERCENT OF APPLIED POWER AND FURNACE TEMPERATURE

Step Number	1	2	3	4	5	6	7	8	9	10	11
Percent of Power	0	10	20	30	40	50	60	70	80	90	100
Temperature °C	20	220	420	620	820	1020	1220	1420	1620	1820	2020

Phase-control circuits have an infinite number of settings. Switching at zero voltage requires a finite number of control settings or steps. There are only eleven steps in the example shown in Fig. 12.1. This may not be adequate for some applications. Table 12.1 shows the percent of power and temperature of an eleven-step system used with a furnace. The range of this furnace is from 20°C to 2020°C.

The furnace temperature could only be set at the eleven values found in Table 12.1. If 902° C were desired, the closest would be 820°C. This is an error of 100°C or 10%.

Increasing t_d increases N_{cs}. Worse inaccuracies may occur with a longer t_d. As an extreme example, let the t_d be 24 hours. Set the furnace for 50% power (1020°C). Full power would be on for 12 hours, then off 12 hours. During the 12 hour on-time, the temperature may reach over 2020°C. During off-time it may cool far below 1020°C.

The system designer must keep N_{cs} high and t_d low. The solution is to decrease source-cycle time (t_s). This usually requires a high-frequency source such as a cycloconverter could provide.

12.2.2 Zero-Voltage Switching Systems

SCRs and triacs are well suited for zero-voltage switching as they inherently commutate (turn off) at zero volts. The more complex task is to trigger (turn on) the SCR at zero volts.

Zero-voltage triggers are available in integrated circuit form from many manufacturers. Two types of *discrete-component* zero-voltage triggers are covered here.

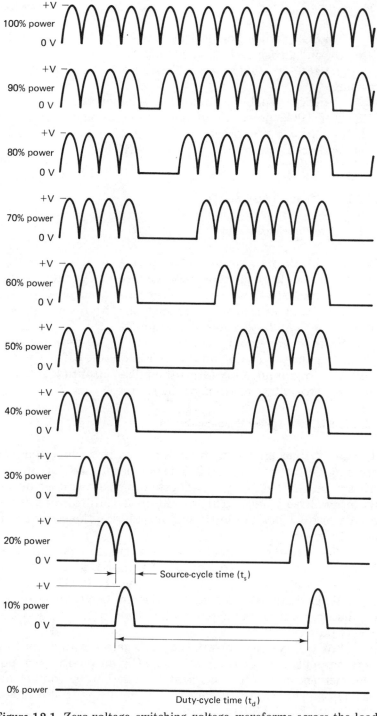

Figure 12.1 Zero-voltage switching voltage waveforms across the load at eleven power settings.

The SCR zero-voltage trigger is made up of two circuits: a *zero-crossing detector*, and a *trigger-enabling circuit*.

The zero-crossing detector furnishes a pulse output at each source-zero crossing.

The trigger-enabling circuit lets selected zero-crossing pulses trigger the SCR. For 100% power all of the zero-crossing detector pulses are enabled. For 50% power the trigger-enabling circuit allows only half of the zero-crossing detector pulses to trigger the SCR.

12.3 DIGITAL ZERO-VOLTAGE SWITCH

12.3.1 Zero-Crossing Detector

Figure 12.2 shows a simple zero-crossing detector. The input to the bridge (V_{ac}) is a sample of the power-source frequency and voltage. The bridge supplies the transistor base with a pulsating dc voltage (V_{BB}) [see Fig. 12.2(c)]. Diode D_1 and capacitor C_1 supply the transistor collector with a fairly ripple-free V_{CC}. D_1 isolates V_{BB} from V_{CC}.

Voltage V_{BB} holds the transistor in saturation except at zero volts. Then, V_{BB} drops to zero and the transistor cuts off placing full V_{CC} across its collector to ground. This forms the pulse output (V_o) [see Fig. 12.2(d)] exactly at zero-voltage crossing of the source.

12.3.2 The Gate-Trigger Enabling Circuit

The gate-trigger enabling circuit is shown in a block diagram with the zero-crossing detector in Fig. 12.3. The enabler is made up of a *duty-cycle oscillator* and an AND gate. The AND gate has a zero-voltage output unless both of its inputs have voltage. This condition is fulfilled only when the duty-cycle oscillator has a positive pulse and only during the instant of the zero-crossing detector pulse.

The zero-cross detector sends a steady train of zero-crossing pulses [V_{ZCD} in Fig. 12.3(c)] to input A of the AND gate. The duty-cycle oscillator sends a rectangular wave [V_{DCO} in Fig. 12.3(d)] to input B of the AND gate. When both voltages are positive the AND gate triggers the SCR. This occurs only during the gate-enabling time of V_{DCO} and then exactly at the V_{ZCD} pulse.

For the example shown, the AND gate has an output for four out of six detector pulses. This turns on the SCR and delivers $\frac{2}{3}$ power to the load. The percent power to the load is varied by a control on the duty-cycle oscillator. This varies the length of the V_{DCO} pulse labeled "gate-enabler time" [see Fig. 12.3(d)].

The duty-cycle oscillator shown in Fig. 12.4 uses a 555 timer integrated

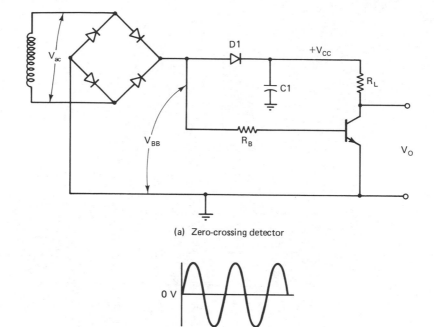

(a) Zero-crossing detector

(b) V_{ac} source voltage

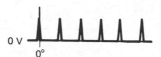

(c) V_{BB} transistor base supply

(d) V_O output pulse every transistor cutoff

Figure 12.2 A simple zero-crossing detector.

circuit. The 555 is available from Signetics Corporation and many other manufacturers. Its duty cycle may be changed by selecting different values of R, R_1, R_2, and C_1. Where $R_t = R + R_1 + R_2$, duty cycle time (t_d) is

$$t_d = 0.685(R_t)C_1 \qquad (12.2)$$

Using a C_1 of 1 μF, t_d is about 692 ms. Using a 60 Hz source, the $t_s = 1/f = 16.7$ ms. This yields an $N_{cs} = t_d/t_s = 41.4$ or about 41 control steps [see Eq. (12.1)]. When the wiper of R is closest to R_2 the gate-enable time (t_{ge}) is longest—about 95%. The wiper closest to R_1 (t_{ge}) is only about 5%.

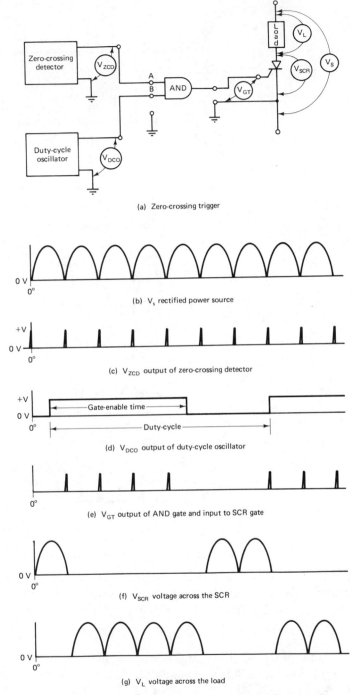

(a) Zero-crossing trigger

(b) V_s rectified power source

(c) V_{ZCD} output of zero-crossing detector

(d) V_{DCO} output of duty-cycle oscillator

(e) V_{GT} output of AND gate and input to SCR gate

(f) V_{SCR} voltage across the SCR

(g) V_L voltage across the load

Figure 12.3 Zero-voltage switching block diagram.

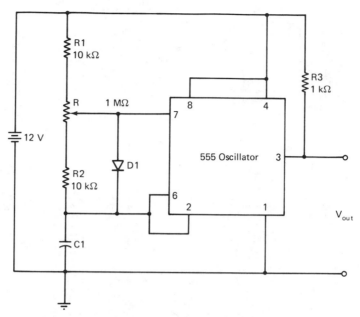

Figure 12.4 The 555 duty-cycle oscillator.

12.4 RAMP ZERO-VOLTAGE SWITCH

12.4.1 The Ramp-Enabler Zero-Voltage Switch

A second method of zero-voltage switching is shown in Figs. 12.5 and 12.6. This method uses a ramp waveform to enable the zero-crossing pulse instead of a duty-cycle oscillator and AND gate. Its shortcoming is a continuous gate drive (see V_G in Fig. 12.5). Yet, similar circuits are still in use in the industry.

This trigger switches the SCR on by switching a transistor off (see Q_1 in Fig. 12.6). When the transistor is off, full-collector supply voltage (V_{CC}) is across its collector to ground. This is connected across the SCR gate-to-cathode and is adequate to fire an SCR.

The transistor must be turned off at the SCR zero-voltage point. The transistor is turned off and on by its base voltage (V_B). V_B is an SCR source-synchronized full wave riding on a variable ramp [see Fig. 12.5(b), (c), and (d)] V_B . The transistor turn-off level (TTOL) is shown by a light line above the zero line. When V_B goes below this line the transistor turns off (see where arrows point from V_B to V_G). Each time the transistor turns off, it turns the SCR on for 180°. In Fig. 12.5 it fires the SCR twice for every cycle of V_s . The SCR delivers one inverter-source V_i cycle to the load through the bridge rectifier. The V_B ramp is set smaller in Fig. 12.5(c). After the transistor is switched on it is held on for three cycles of V_i. This

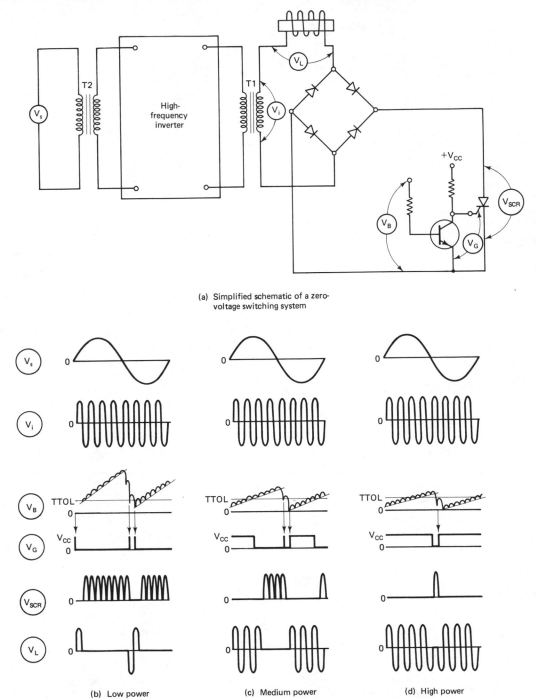

(a) Simplified schematic of a zero-voltage switching system

(b) Low power   (c) Medium power   (d) High power

Figure 12.5 Simplified schematic and waveforms of zero-voltage switch controlling an induction furnace.

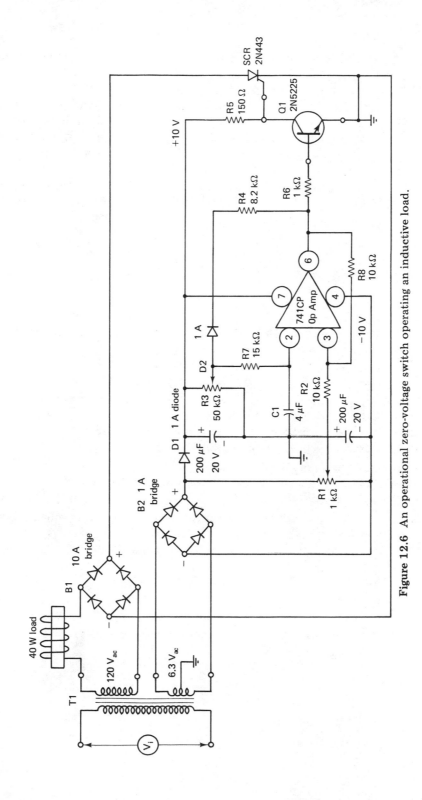

Figure 12.6 An operational zero-voltage switch operating an inductive load.

enables the SCR at five consecutive V_i zeros. Figure 12.5(d) shows the V_B ramp set lower yet. The transistor is off nearly all of the time, enabling the SCR to deliver nearly full power to the load.

Figure 12.6 shows the complete zero-voltage switching circuit. It includes the V_B source not shown in Fig. 12.5. Potentiometer R_1 provides the full-wave rectified portion of V_B to the operational amplifier. Potentiometer R_3 provides the variable ramp. R_1 is initially set, then R_3 is used to vary power to the load.

The V_B source may be used to drive two or more transistor/SCR combinations. This could be used to trigger any of the SCR/load combinations shown in Figs. 9.6 or 10.7.

PROBLEMS

12.1. Zero-voltage switching is used when a combination of two types of applications are necessary. What are these two types?

12.2. What are the two dangers of excessive RFI?

12.3. What type of SCR control eliminates excessive RFI?

12.4. What specific application uses zero-voltage switching?

12.5. What specific type of circuitry is used to commutate (turn off) the SCR at zero voltage? Explain your answer.

12.6. What device is used to turn on an SCR at zero voltage?

12.7. What does R_1 in Fig. 12.6 control?

12.8. What does R_3 in Fig. 12.6 control?

12.9. What is the voltage across the off transistor in Fig. 12.6?

12.10. What is the approximate voltage across an on transistor with a V_{cc} of 20 V?

12.11. Find the number of control steps in a zero-voltage system with a duty-cycle time of 60 ms and a source frequency of 1000 Hz.

12.12. List the two circuits used in a zero-voltage trigger.

12.13. What is the function of D_1 in Fig. 12.2(a)?

12.14. Find the duty-cycle time of the oscillator shown in Fig. 12.4 when R_t is 500 kΩ and C_1 is 0.1 μF.

12.15. Why are SCRs inherently suited for zero-voltage switching applications?

12.16. What is the advantage of the rectangular enabler circuit shown in Fig. 12.3 over the ramp-enabler trigger in Fig. 12.6?

Static Switching and Electronic Time Delays

13

13.1 STATIC-SWITCHING APPLICATIONS

Electronic static-switching applications consist of circuits which replace manual on-off switches, relays, circuit breakers, fuses, and flashers. Some commonplace uses are switching on emergency lighting during blackouts and in flashers on automobiles and portable roadside warning lights.

These switches can easily be actuated automatically. They are inexpensive and reliable. They may be used to actuate either ac or dc circuits. The term static means "not changing." A static switch turns a power off or on, but does not vary it in other ways. The term static excludes proportional control, where the percentage of off or on is varied. Even in the case of flashers the on-off changes are constant.

13.2 CROWBAR CIRCUITS

13.2.1 Electronic Crowbar Circuits

Crowbar circuits are current and/or voltage overload circuits which protect a load by disconnecting it from the line. The term crowbar stems from the act of prying or disconnecting a circuit from the line. Critical loads are often threatened by variations in source voltage. The primary voltage source of industry is electrical power company service. This service often has inadequate voltage regulation. This primary service reflects on any unregulated power derived from it. The crowbar overvoltage-protection circuit keeps overvoltage from destroying a load.

Certain loads may draw more current than desirable, even at rated voltage. For example, a motor, heater, or transformer may have a few shorted windings lowering its total resistance. Then, it would draw too much current and may destroy the remaining winding. A battery charger may inherit a severely discharged battery. This may cause the charger to burn out. The crowbar overcurrent-protection circuit keeps overcurrent from destroying a load or load-servicing device.

Figure 13.1 shows a basic crowbar circuit. The overload detection circuits are omitted for simplicity. They are shown in Figs. 13.2, 13.3, and 13.4. A signal from either overload circuit fires the SCR. This energizes the overload (OL) relay. The normally-closed OL contacts open and disconnect the load from the source.

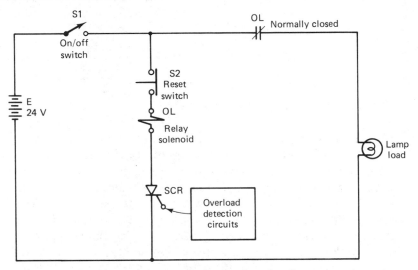

Figure 13.1 Basic crowbar circuit.

13.2.2 The Crowbar Overvoltage-Protection Circuit

A crowbar overvoltage-protection circuit is shown in Fig. 13.2. When the source voltage involuntarily increases to a value unsafe for the load, the UJT triggers. This fires the SCR and opens the circuit. This circuit may be adapted for ac sources.

Once the SCR is fired, it latches on. This holds the relay on, and its OL contacts off. The operator must correct the problem and then push the S_2 reset switch to turn the circuit back on.

The value of overload voltage which turns the load off may be set using R_2. To set R_2, move its wiper toward the bottom of R_2. Next, connect a

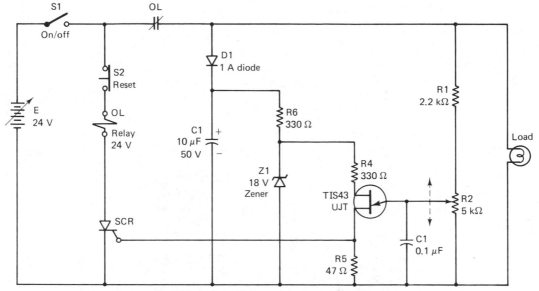

Figure 13.2 Crowbar overvoltage-protection circuit.

variable source in place of E and adjust it to the desired overvoltage. Next, slowly raise the R_2 wiper until the OL contacts open.

The circuit may be placed across any nominal 24 V source. If an overvoltage should occur, the voltage at the UJT emitter triggers it. The output from the UJT fires the SCR. The SCR energizes the OL relay solenoid. This opens the OL contacts and turns off the power to the load.

Diodes D_1, C_1, R_6, and Z_1 provide an 18 V regulated supply for the UJT base. Resistors R_1 and R_2 supply the UJT emitter with a sample of the source voltage. When the source voltage is too high the UJT becomes forward-biased from emitter to base 1. This fires the UJT and the emitter-to-base 1 resistance becomes a short. A current surge flows from the emitter to base 1 and through R_5. The surge through R_5 causes a voltage pulse across it. This pulse fires the SCR.

13.2.3 The Crowbar Overcurrent-Protection Circuit

Figure 13.3 shows a crowbar overcurrent-protection circuit. The R_9 wiper is set just below the value of voltage required to fire the UJT with rated current through the load. Any load current must flow through R_{11}. Excess load causes a higher voltage crop across R_{11}. This increases the voltage at point Ⓐ and point Ⓦ with respect to ground. The emitter of the UJT becomes forward-biased with respect to the eta point (see Sec. 5.3). This triggers the UJT causing a high current through R_{10}. This surge through R_{10} causes a voltage pulse to be developed across it. The pulse fires the SCR and energizes

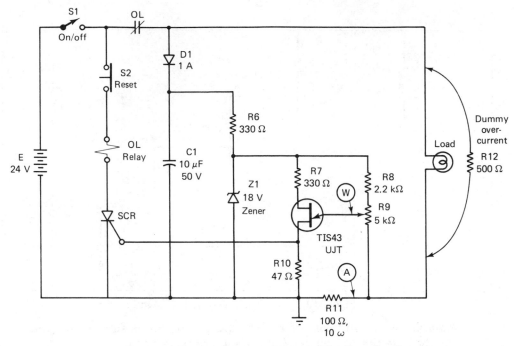

Figure 13.3 Crowbar overcurrent-protection circuit.

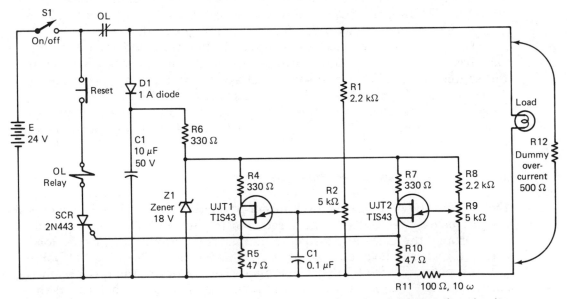

Figure 13.4 Crowbar overvoltage- and overcurrent-protection circuit.

the OL relay, disconnecting the load. The basic crowbar circuit latches the load off until an operator pushes the S_2 reset.

13.2.4 The Crowbar Overvoltage- and Overcurrent-Protection Circuit

The crowbar overvoltage- and overcurrent-protection circuit is shown in Fig. 13.4. If either the source voltage or the load current exceeds the normal value, the circuit is turned off. Then, the operator may correct the problem and reset the circuit.

The circuit works exactly like a combination of Figs. 13.2 and 13.3. Both UJT sensing circuits use the Zener regulated source and their outputs are connected together. If either the overvoltage or the overcurrent sensors trigger, or both trigger, the SCR fires. Then, the overload relay disconnects the load from the source.

13.3 FLASHERS

13.3.1 Lamp Flashers

Lamp flashers are low-power inverters with the specific application of blinking or flashing lamps.

Figure 13.5 is a popular flasher circuit. A UJT oscillator supplies triggers to both SCRs. SCR_2 drives the lamp. SCR_1 and C_1 commutate SCR_2. The UJT oscillator consists of R_1, C_1, R_2, R_3, R_4, and the UJT (see Sec. 5.3).

Figure 13.6 shows the operation of the flasher. Figure 13.6(a) shows the basic flasher. The initial UJT pulse reaches SCR_1 first. C_3 and R_6 slightly delay the pulse to SCR_2. Thus, SCR_1 fires and C_2 charges to E

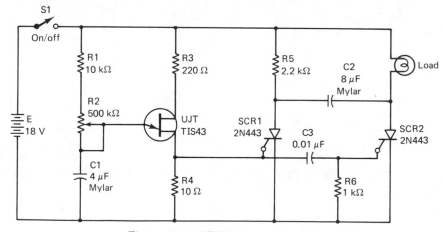

Figure 13.5 UJT/SCR lamp flasher.

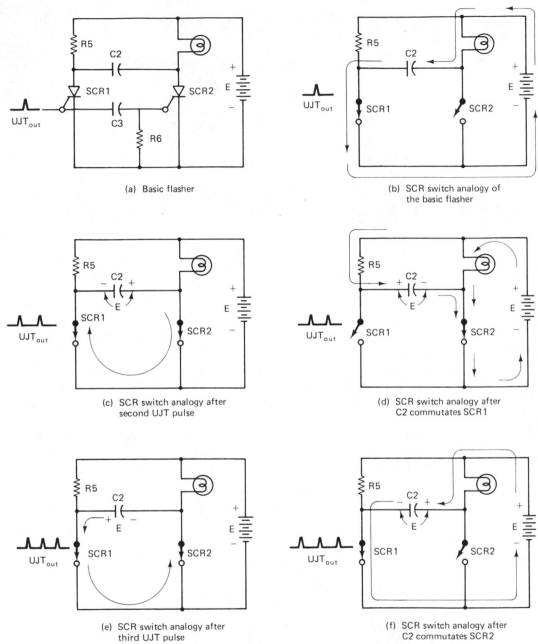

(a) Basic flasher

(b) SCR switch analogy of the basic flasher

(c) SCR switch analogy after second UJT pulse

(d) SCR switch analogy after C2 commutates SCR1

(e) SCR switch analogy after third UJT pulse

(f) SCR switch analogy after C2 commutates SCR2

Figure 13.6 The operation of the flasher shown in Fig. 13.5.

through the lamp [see Fig. 13.6(b)]. C_2 is charged to E when the second pulse reaches SCR_1 and SCR_2 and SCR_2 fires [see Fig. 13.6(c)]. Both SCRs are on allowing C_2 to discharge. The C_2 discharge current holds SCR_2 on, but commutates SCR_1 [see Fig. 13.6(c)]. Figure 13.6(d) shows SCR_1 off and SCR_2 on. Current from E lights the lamp and at the same time charges C_2 through R_5 and SCR_2. This time C_2 charges in the opposite direction [see Fig. 13.6(d)]. The third UJT pulse fires SCR_1 and both SCRs are on. Then, C_2 discharges through both SCRs holding SCR_1 on and commutating SCR_2 [see Fig. 13.6(e)]. Figure 13.6(f) is back at the same state as shown in Fig. 13.6(b), where events repeat on consecutive UJT pulses. The lamp flashes on when SCR_2 is on, and off when SCR_2 is off.

13.4 TIME DELAYS

13.4.1 UJT/SCR Time-Delay Circuit

Figure 13.7 shows a UJT/SCR time-delay circuit. After S_1 is turned on a delay of from 20 ms to 10 min may be made to occur before the load turns on.

The delay is based on the RC time constant of $R_T + R_1$ and C_1. It takes about one time constant for C_1 to charge to the firing point of the UJT. Thus, the longest delay for the values shown in Fig. 13.7 is 3 s (see Sec. 5.3). After the UJT fires, the current surge through R_L provides a pulse which turns on the SCR. This turns on the load. The load remains on until power is turned off via S_1. After S_1 is turned off, C_1 discharges through the UJT and R_L, and remains ready for the next delay.

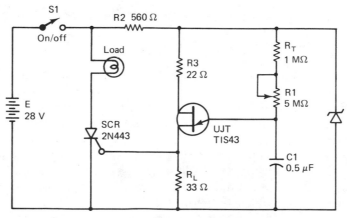

Figure 13.7 UJT/SCR time-delay circuit.

13.4.2 Testing the PUT

The PUT was briefly described in Sec. 6.2. It may also be thought of as a complementary SCR. It may be tested with an ohmmeter as an SCR, except its gate must be shorted to its cathode instead of its anode in order for it to turn on (see Sec. 6.1.2). The PUT turns on when its anode-to-cathode is forward-biased and its gate is greater than 0.7 V more negative than its anode.

13.4.3 The PUT Time Delay

Figure 13.8 shows a PUT/SCR time delay. When S_1 is off C_1 is shorted by R_2 and has zero volts across it. When S_1 is turned on C_1 is no longer shorted by R_2. C_1 charges through R_5 and R_1. When its change gets to 0.7 V it fires the PUT. This sends a surge of current through R_4. The surge provides a pulse of voltage which fires the SCR and turns on the load.

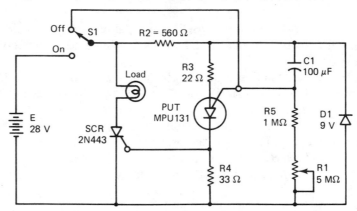

Figure 13.8 PUT/SCR time-delay circuit.

13.4.4 The Sequential Time Delay

Figure 13.9 shows a PUT/SCR time delay which turns on three lamps in sequence. When S_2 is pushed on lamp L_1 turns on. After C_4 charges up to a voltage 0.7 V greater than the gate, lamp L_2 turns on. SCR_3 turns on lamp L_2 and energizes the second PUT time delay. Then, C_5 charges up to 0.7 V greater than the gate of PUT_2 firing PUT_2. This turns on SCR_4 and lamp L_3. All three lamps remain on until S_2 is released, turning circuit power off.

The PUT time delay circuit(s) used in Fig. 13.9 shows why the PUT is described as programmable. The values of the voltage divider R_9 and R_{10} program the voltage at which PUT_1 fires. By making R_9 larger, C_4 charges quicker to 0.7 V greater than the gate. Thus, the time delay is less. In a like manner, R_{12} and R_{13} program PUT_2.

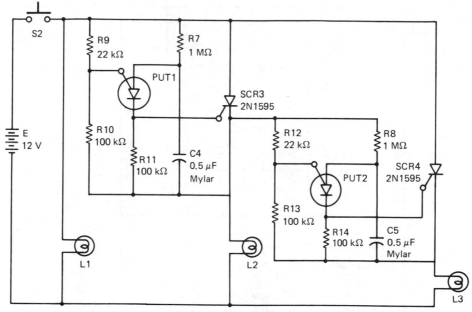

Push-button-held open switch used in place of
on/off flasher in Fig. 13.5

Figure 13.9 PUT/SCR switch sequentially turning three lamps on.

13.4.5 The Sequential Flasher

Figure 13.10 is a sequential flasher. After it is turned on with S_1, L_1 lights and remains on, then L_2 lights and remains on, finally L_3 lights. Next, all three lamps go off. Then, the procedure repeats itself. A similar circuit is used on the turning-signal tail lights of the Mercury Cougar built by Ford Motor Company.

The PUT/SCR sequential switch in Fig. 13.9 is driven by the flasher shown in Fig. 13.5. The period the flasher is on must be long enough to provide time for both PUT time delays. Or conversely, both PUT time delays must be adjusted to fire during the time the flasher is on.

PROBLEMS

13.1. List five devices static-switching circuits replace.

13.2. List two types of crowbar circuits.

13.3. What two devices in the circuit in Fig. 13.5 commutate SCR_2?

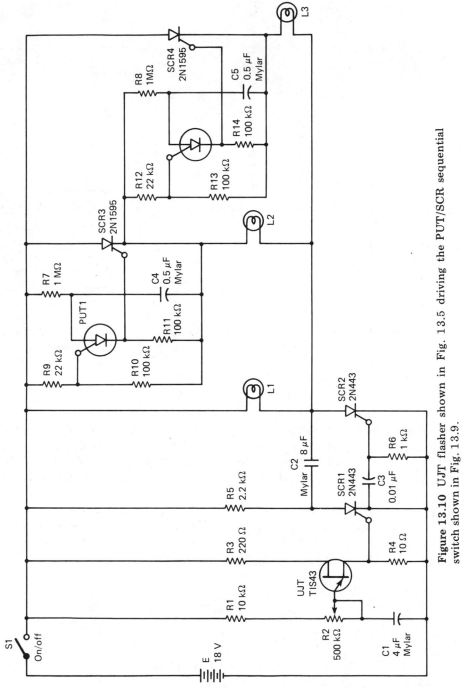

Figure 13.10 UJT flasher shown in Fig. 13.5 driving the PUT/SCR sequential switch shown in Fig. 13.9.

13.4. What device is C_2 in Fig. 13.5 connected directly across when SCR_2 is on?

13.5. List two components which will increase the time delay of the circuit shown in Fig. 13.7 if they are increased in value.

13.6. List the component(s) of the discharge path of C_1 in Fig. 13.8.

13.7. How does increasing R_9 or R_{12} affect the time delay in the circuit shown in Fig. 13.9?

13.8. How does increasing R_{10} or R_{13} affect the time delay in the circuit shown in Fig. 13.9?

13.9. How does increasing the value of C_4 affect the time delay in the circuit of PUT_1 in Fig. 13.9?

13.10. Which SCR turns off the lamps in Fig. 13.10 when it turns on?

Triac/Diac Circuits

14

14.1 INTRODUCTION

Triac/diac circuits dominate the consumer market in lamp dimmer and variable-speed electric drills. These circuits are also replacing SCRs in many industrial applications. The main limitation on the triac is that they are not available with an $I_{T(\text{rms})}$ current rating over 25 A.

The triac was briefly covered in Sec. 6.7 and the diac in Sec. 4.2.1. The schematics, characteristic curves, and equivalent circuit of the triac and diac were shown in Figs. 4.5, 4.6, and 6.10.

14.2 THE DIAC

The diac is a bidirectional diode. It operates like two diodes in series. Its two leads are interchangeable. The turn-on voltage of most diacs is about 20 V. With 20 V across a diac it conducts and acts like a low resistance with about a 3 V drop across it. At circuit voltages less than 20 V it acts like an open circuit.

Figure 14.1 shows how a diac may be tested in a dc circuit. The circuit consists of a zero to 30-V variable dc source, a 10-kΩ resistor, a zero to 10-mA ammeter, and a zero to 50-V_{dc} voltmeter. Connect the circuit with the dc source set at zero. Then, slowly increase the voltage. The voltage across the diac should rise, but the current should be zero, until about 20 V. At 20 V the voltage should drop to between 3 and 5 V, while the current should rise sharply to about 5 mA.

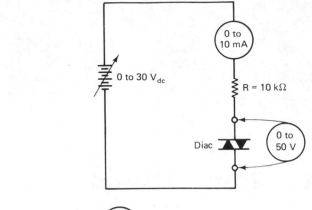

Figure 14.1 Diac test circuit.

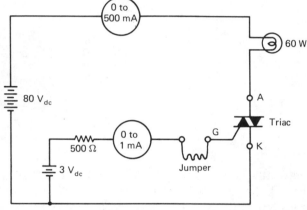

Figure 14.2 Triac test circuit.

14.3 TRIAC FIRING CHARACTERISTICS

A triac test circuit is shown in Fig. 14.2. Set the 500-Ω potentiometer to maximum resistance. Connect the triac in the circuit and the gate jumper. Then, slowly decrease the 500-Ω potentiometer, increasing gate current until the triac fires and lights the lamp. The lamp should remain on after the gate jumper is removed.

The triac should fire in this same manner with either or both dc sources reversed. That is, it should fire with a positive anode and a positive gate, a positive anode and a negative gate, a negative anode and a negative gate, or a negative anode and a positive gate. The SCR-equivalent circuit of a triac (see Fig. 6.10) is not truly accurate. A better equivalent circuit of a triac is two SCS's connected in parallel, anode to cathode, along with all four gates connected together.

14.4 THE BASIC TRIAC/DIAC CONTROL CIRCUIT

The basic triac/diac control circuit is shown in Fig. 14.3. Varying R_2 varies the time C_1 charges to the diac trigger voltage. When the diac turns on it fires the triac. This turns on the load until the source voltage (V_S) goes

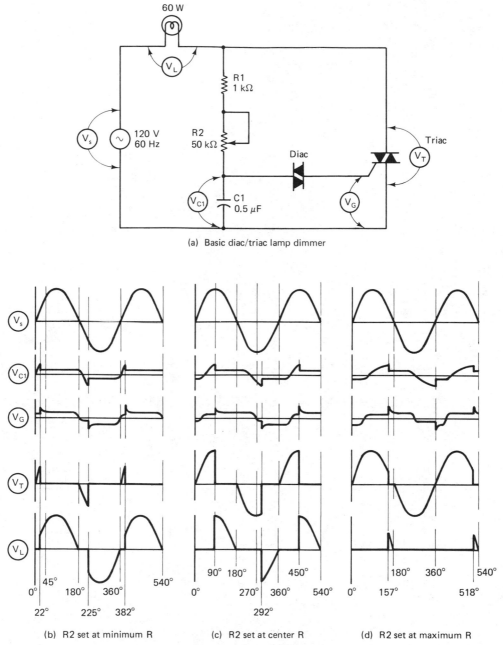

60 W

V_L

R1
1 kΩ

V_s 120 V
60 Hz

R2
50 kΩ

Diac

Triac

V_T

V_{C1} C1
0.5 μF

V_G

(a) Basic diac/triac lamp dimmer

V_s

V_{C1}

V_G

V_T

V_L

45°
0°   180°   360°   540°
22°   225° 382°

90° 180°   450°
0°   270° 360°   540°
292°

180°   360°   540°
0°   157°   518°

(b) R2 set at minimum R   (c) R2 set at center R   (d) R2 set at maximum R

Figure 14.3 A basic triac/diac lamp dimmer and waveforms.

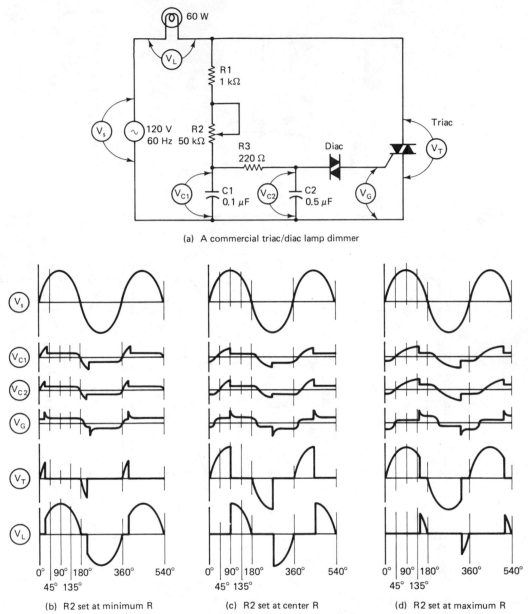

(a) A commercial triac/diac lamp dimmer

(b) R2 set at minimum R   (c) R2 set at center R   (d) R2 set at maximum R

Figure 14.4 A commercial triac/diac lamp dimmer and waveforms.

through zero. At zero volts the triac turns off. It remains off until the capacitor C_1 charges to the diac's trigger voltage again.

Potentiometers R_2 and C_1 are selected for an RC time constant capable of turning on the diac/triac for nearly any conduction angle from $0°$ to $180°$ (see Fig. 7.4). R_1 is a protection resistor so that the diac and triac gate are never exposed to full-line voltage.

Figure 14.3(b) shows the circuit waveforms with R_2 set at minimum resistance. The voltage across C_1 reaches the diac trigger (turn-on) voltage at $22°$ after the source voltage goes positive through zero. The triac latches on early during each half cycle, delivering nearly full power to the load.

Figure 14.3(c) shows the waveforms with R_2 set so that the triac delivers half power to the load. Figure 14.3(d) shows the waveforms with R_2 set at maximum resistance. Here, the RC time constant is longest. It takes nearly the whole half cycle before C_1 charges to the diac trigger voltage. Thus, the triac delivers low power to the load.

Figure 14.3 does not produce symmetrical load voltages. The time the negative load voltage is on is less than the time the positive is on [see V_L Fig. 14.3.(a), (b), and (c)]. Some unsymmetry is caused by the triac firing characteristics, but most of it is due to the *hysteresis* caused by the capacitor. The capacitor will retain some charge in the polarity of the initial voltage placed across it. In Fig. 14.3 the waveforms show that C_1 favors the positive charge. This delays the negative charge. This unsymmetry is corrected by the addition of R_3 and C_2 shown in Fig. 14.4.

The circuit shown in Fig. 14.4 is popularly used in many lamp dimmers sold commercially.

PROBLEMS

14.1. What components are most popular in commercial lamp dimmers and variable-speed drills?

14.2. What is the $I_{T\,(\mathrm{rms})}$ rating of the largest triac manufactured at the time of this printing?

14.3. What is the equivalent circuit of the diac?

14.4. What is the approximate trigger voltage of most diacs?

14.5. What is the approximate voltage drop across an ON diac?

14.6. Sketch the SCS-equivalent circuit of the triac.

14.7. List the two combinations of anode-gate voltage polarities that are not shown by the SCR-equivalent circuit of the triac.

14.8. In order to deliver full power to the load in Fig. 14.3, at what value of resistance should R_2 be set?

14.9. Express the relation of V_S, V_T, and V_L with an algebraic equation (see either Fig. 14.3 or 14.4).

14.10. Describe the function of R_3 and C_2 in Fig. 14.4.

Index